Increasing Productivity and Profit through Health & Safety

The Financial Returns from a
Safe Working Environment
2nd Edition

Increasing Productivity and Profit through Health & Safety

The Financial Returns from a
Safe Working Environment
2nd Edition

Maurice Oxenburgh, Pepe Marlow and Andrew Oxenburgh

CRC PRESS

Boca Raton   London   New York   Washington, D.C.

Library of Congress Cataloging-in-Publication Data

Oxenburgh, Maurice, 1935–
    Increasing productivity and profit through health and safety: the.
financial returns from a safe working environment / Maurice Oxenburgh,
Pepe Marlow & Andrew Oxenburgh.
        p. cm.
    Includes bibliographical references and index.
    ISBN 0-415-24331-9 (hardback)
    1. Industrial hygiene – Economic aspects. 2. Industrial safety –
Economic aspects. 3. Industrial productivity – Computer programs.
I. Marlow, Pepe, 1961– II. Oxenburgh, Andrew, 1965– III. Title.

HD7261 .O94 2003
658.3′82—dc21   2002155081

British Library Cataloguing in Publication Data

A catalogue record for this book is available from the British Library

Visit the CRC Press Web site at www.crcpress.com

© 2004 Maurice Oxenburgh, Pepe Marlow & Andrew Oxenburgh

No claim to original U.S. Government works
International Standard Book Number 0-415-24331-9
Library of Congress Card Number 2002155081
Printed in the United States of America 1 2 3 4 5 6 7 8 9 0
Printed on acid-free paper

Contents

About the authors

Maurice Oxenburgh

Maurice Oxenburgh graduated from the University of New South Wales with a doctorate in biochemistry but, for the past quarter of a century, has worked in occupational health and safety. His principal interests were originally in occupational hygiene, mostly noise control and ventilation, but more recently he has added ergonomics to his slate, mostly in prevention of upper limb injuries and in workplace design. He has served as an expert witness in common law personal injury matters and in March 2000 he gave evidence to support OSHA's proposed 'Ergonomics Standard' in Washington, DC.

Whilst working in industry, and later as a consultant to industry, he realised that although managers wanted efficient workplaces they only saw safety as a cost. His experience showed otherwise and he has taken it as an article of faith that a safe workplace is more effective, efficient and productive than one that is not safe. This is his second book on this theme.

Dr Oxenburgh is presently ensconced as Emeritus Research Scholar at the National Institute for Working Life (Sweden) continuing his work on developing methods for measuring worker safety and productivity. He is a Fellow of the Ergonomics Society of Australia.

Pepe Marlow

Pepe Marlow has an undergraduate qualification in Physiotherapy and a postgraduate qualification in Economics. Since 1997 Pepe has worked as a consultant specialising in short-term projects in the occupational health, safety and injury fields. Her clients include both private and public organisations – individual employers, insurers, rehabilitation providers and the New South Wales WorkCover Authority.

Prior to 1997 she worked with the National Occupational Health and Safety Commission (Australia), where she was Manager of Standards Development and Planning.

One focus of her work has been analysis and demonstration of the cost-benefit of occupational health and safety. While at the National Occupational Health and Safety Commission Pepe was responsible for oversight of contracts for the analysis of the cost-benefit of proposed new occupational health and safety standards. Since working as a consultant Pepe has worked with Maurice Oxenburgh to develop a series of cost-benefit case studies.

Andrew Oxenburgh

Andrew Oxenburgh is a Bachelor of Applied Science in Computing Science from the University of Technology, Sydney.

During the past ten years he has gathered extensive experience in the banking and financial services industries developing financial and stock handling programs and web sites. Currently he is consulting to an international financial services company in North West England.

He enjoys skiing, diving, travelling, though not all at once, and often with more enthusiasm than skill.

Foreword

Precor is a medium-sized US company specialising in the manufacture of premium quality fitness equipment for home and fitness facility use. We are firmly committed to excellence and leadership in sound environmental, health and safety practices as a *routine* part of our business pursuits. To this end, all economically and technically reasonable opportunities for ensuring public and employee safety, hazardous waste reduction, recycling and regulatory compliance methods are pursued with the same energy and commitment Precor displays toward its high quality business strategies.

Dedicated personnel using proactive systems and policies manage and monitor all aspects of environmental protection, employee health and safety and regulatory compliance striving for continuous improvement.

Three years ago we started a new process whereby 'cell' teams were formed throughout our assembly areas. The primary objective of each cell team was to solicit ideas from our assembly level folks and track/implement them on a large scale. What happened quickly was a system of idea generation and quick follow-up that snowballed into a machine that drove down assembly times and resulted in an unprecedented willingness to contribute by all people. The cell meetings in themselves have been conducted in a way that everyone feels at ease with sharing. They feel rewarded when they perform the change themselves or they see the change happen at someone else's hand, because it was their idea. We also do a load of spot rewarding with Seattle Mariners baseball tickets. The number of safety ideas and the immediate support and resolution of those ones provide the real heart of the cell process. I believe that as long as we support cell meetings, fund the changes, and continue the 'every idea counts' in the cell meetings, the idea mill will continue to 'make hay' for us.

Fixing safety issues has people understanding that we really do care about them and put our money on the table to prove it. When we ask for ideas to reduce assembly time (and ultimately use fewer people) they are right with us and contribute freely. We share data with them about profits, production numbers, what using fewer people to do the job means to them, and all of the data used to fuel the business. We have been very upfront about 'doing more with less' and people appreciate the honesty. We consider 'safety' and 'ergonomics' as team sports and support them to encourage participation.

Without these people contributing at the assembly level of our manufacturing business we would not have increased our business. In three years we increased productivity per employee by over twice, along with increased sales. At the same time our injury record has decreased from 5.9 injuries per 100,000 worked hours to 1.7. This has reduced our workers' compensation bill from $0.17 per paid hour to less than $0.05.

You can see that safety pays at Precor and when you include that whole picture of empowerment, committing resources, and follow-up with an attractive recognition program, it does more. Support of the safety effort helps to grasp the essence of a healthy working environment that fuels future company growth!

Bill Setter
Director of Operations
Precor Commercial Division
Seattle, Washington, USA
www.precor.com

Acknowledgements

Professor Åsa Kilbom (National Institute for Working Life, Stockholm) was the inspiration and stimulus for the first edition of this book and the spur to update it. We wish to record our appreciation not only for her inspiration to these present authors but for her leadership in ergonomics and particularly in upper limb injury prevention during the past two or so decades.

Maurice Oxenburgh wishes to acknowledge and thank Professors Christer Högstedt and Ewa Wigaeus-Tornqvist of the Ergonomics Program of the National Institute for Working Life, Stockholm, for financial support and encouragement during the course of writing this book.

In particular Pepe Marlow wishes to thank her father and mother Jim and Peggy Marlow, husband Kevin Shaw and sons Christopher and Michael for their continued encouragement and support. Although Peggy did not live to see the final result of this project she was supportive of her daughter at all times.

We thank Tony Moore of Taylor & Francis for his encouragement and understanding when, for personal reasons, we were not able to keep to our deadlines.

We also have special thanks to all those who have given us so much of their time when we visited their workplaces. Some of the case studies have been reproduced from the first edition but most are new, and we wish to thank: Dr Denis D'Auria, Consultant Occupational Physician and Director, Occupational Health Services, Barts and The London NHS Trust, London, UK; Mr David Armstrong, President, NSW Welders Association, NSW, Australia; Mr Michael Bishop, General Manager, BITE Teleperformance, Göteborg, Sweden; Mr Richard Connett, General Manager, Occupational Health Services, Barts and The London NHS Trust, London, UK; Mr Trevor Deeming, Manager, Liverpool Call Centre, Centrelink, NSW, Australia; Mr Joe Donachie, Longwall Superintendent, Baal Bone Colliery, NSW, Australia; Mr Ted Greenhill, Cleaning Co-ordinator, University of New South Wales, Australia; Mr Robert Pettersson, Works Manager, Bromma Conquip, Vällingby, Sweden; Mr Barry Phillips, Mr Mark Phillips, Newtech Tube Co., Kurnell, NSW, Australia; Mr Philip Robinson, J. Sainsbury plc, UK; Scottish Courage, Brewing Ltd, UK.

Much of the material in Chapter 7 was prepared by Maurice Oxenburgh for the Ron Cumming Memorial Lecture at the Ergonomics Society of Australia's Scientific Meeting, November 2001.

Some of the information in Chapters 5A and 5C is reproduced by permission of WorkCover Authority of NSW from its booklet *Linking Productivity and OHS: a Quick Guide to Costs and Benefits* (1999), copies of which may be obtained from WorkCover, GPO Box 5368, Sydney, NSW, 2001, Australia, quoting catalogue number 753.

We wish to thank our friends and colleagues who gave their advice during the preparation of this edition; in particular: Guy Ahonen, Chris Aickin, Tomas Berns, Verna Blewett, James Bowden, John Burton, David Caple, Philip Cohen, John Danielsson, Jögen Eklund, Kaj Frick, Anders Grönberg, Carina Haglund, Roger Hall, Ebba Hamelberg, Olle Hammerström, Hal Hendrick, Wendy Ilford, Scott Jensen, Marilyn Jukes, Danny Keeble, Aidre Long, Marcia Lusted, Elaine Mitchell, Chris Neville, Robert Pettersson, Tomas Rogalin, Bev Savage, Robert Scott, Kevin Shaw, Ray Simons, Russell Stevens, Per Tengblad, Allan Toomingas, William Warner, Peter Westerholm, Ian Winham, Anne Wyatt.

Maurice Oxenburgh, Stockholm
Pepe Marlow, Sydney
Andrew Oxenburgh, Manchester
June 2003

1 Introduction

Contents

1.1 More ergonomics!

What! another book on ergonomics? After all, there are already many excellent texts on ergonomics, for both the specialist and the non-specialist, so is this book any different? Well, yes. This is not a book about solving ergonomics problems; it is a book to encourage expenditure on ergonomics solutions by those who hold the purse strings. The methodology for these solutions is through cost-benefit analysis; after all, '*good ergonomics is good economics*'.

When we are asked by friends and others not connected with occupational health and safety what book are we writing, the reply is that the book is to show that good working conditions are good for business. Inevitably they look astonished and say 'well of course that is so'. This book sets out to explain what the layperson knows intuitively; that better working conditions lead to increased safety and higher productivity and profit.

Primarily this book is for those people who deal directly with people at work: human resource and personnel departments, senior management, middle management and supervisors. But, as we all want to get our ideas accepted, ergonomists and other occupational health and safety practitioners should know how to put their ideas into an economics and productivity context as well as an injury prevention context – this book is for you, too. To put it crudely, money talks and we, the practitioners, at times need to use the same language.

As eventually all work comes down to people, even those whose work is seemingly remote from people, for example accountants, will find that they need to look more closely at the people side of business – people are an asset, not just a cost.

In many ways this book is an update of the first edition, (Oxenburgh, 1991), which was written at the beginning of the 1990s. The objective of the first edition was to give many case studies (over 60) to show that ergonomics and occupational health and safety can improve the working conditions of employed people without reducing the profitability of the enterprise. A cost-benefit analysis was derived for about half of these cases, and the present book contains a more detailed analysis of the costs for implementing ergonomics solutions and for quantifying the benefits than was possible previously. The costs derive from the implementation of better working conditions while the benefits come from a safer workplace with lower injury costs, higher productivity, reduced warranty costs, improved staff retention, multi-skilling and so on.

In the first edition a generic tool, the Productivity Model, was developed for cost-benefit analysis. This model was based on ideas originally developed by Dr Paula Liukkonen of Stockholm University. The paper form of the model, included with the book, was subsequently produced as a computer program and a basic version of this software, productAbilityBasic, is included with the present edition.

1.2 The objective of this book

The objective is simple: to improve the working conditions of employed people including safety, training, lack of discrimination and other aspects that lead to a pleasant and healthy life at work. The methodology is through showing that productivity and profit for the enterprise are not incompatible with good working conditions for its employees.

In 1959 the International Labour Conference adopted Recommendation No. 112, which defined the purposes of an occupational health service as:

- protecting the workers against any health hazard which may arise out of their work or the conditions in which it is carried on
- contributing towards the workers' physical and mental adjustment, in particular by the adaptation of the work to the workers and their assignment to jobs for which they are suited
- contributing to the establishment and maintenance of the highest possible degree of physical and mental well-being of the workers.

These are still valid aims.

We will not be discussing wage issues *per se*, although wages and variation in wages are factors used in cost-benefit analysis. Our experience is that there is not necessarily a link between safe working conditions and remuneration. One of the first tasks one of the authors (Maurice Oxenburgh) had when he entered the field of occupational health and safety was to dissuade managers and unions from the view that it was quite satisfactory to have dusty and dirty working conditions for labourers as long as you paid them 'dirt money' in recompense. He actually won that battle.

What the authors of this book propose is a system to encourage management to assess their workers more carefully, particularly with regard to their value to the enterprise. This is not a management tool in isolation but a supplement to other management tools. Moreover, we are restricting our view to what an enterprise may do (microeconomics) rather than the wider society (macroeconomics).

Many of the accounting tools used to manage workplaces are based on legislation, taxation and/or cost reduction and thus only measure expense. These systems ask questions such as: can we reduce our tax burden? Can we out-source a function of our business and reduce costs? These are essential questions but do not address productivity and profitability, only potential loss. The objective of this book is more optimistic: it is to enable management to understand clearly that good working conditions lead to improved productivity and profit. '*Good ergonomics is good economics.*'

Even when safety and risk management audits identify workplace hazards, the bare cost of correction may deter management from making the necessary finances and resources available. Identifying a potential problem or hazard may be accompanied by cost avoidance (doing nothing) rather than the benefits of doing something. The value of a cost-benefit analysis is that it is a tool to ensure that the benefits of an intervention are measured in the same way as the costs and overcome the cost avoidance impulse.

Quite unashamedly this book, as did its predecessor, accentuates the role of the worker. There is a large body of literature for measuring the cost of the non-worker side: energy, raw materials, rent, maintenance, sales, communication and so on. We do not deal principally with that side of industry, although the occupational health and safety practitioner must be aware of these costs in discussions when requesting resources for a safety program. These aspects are further discussed in Chapters 2 and 4.

What this book does *not* say is that cost-benefit is the only tool to be used or, even more importantly, that if a project is not cost-effective it need not go ahead; this latter point is discussed in Chapter 7. Cost-benefit analysis is simply a tool to assist in the implementation of occupational health, safety and ergonomics interventions for better working conditions. There are many tools and this is just one of them, although we do believe that, when used correctly, it is a powerful tool.

1.3 Cost-benefit analysis models for occupational health and safety

This book essentially addresses 'how to encourage management to put in place better working conditions through using economics arguments'. No claim is made that this is a new approach, as others have trodden this path before. The following is a brief synopsis of some of the steps taken over the past few years.

From the mid-1970s, Professor Nigel Corlett (Corlett, 1976) discussed cost and benefits from human resources studies, with a seminal paper in 1988 (Corlett, 1988). Discussing various parameters for such a costing, he noted:

> In the nature of things any evaluation is only partial. This can be from inadequacies and insufficiencies in models, in variables and in methods. What is unnecessary is that they should be deliberately insufficient in their perspective. Such a state can arise when models are imposed on situations, not adapted to the cases under investigation. To leave cost-benefit analysis to accountants, to economists, even to ergonomists, is to build in a cause for failure. The utility of professionals, particularly in combination, is undoubted but the understanding of what has to be measured can come only with the aid of those involved. This applies particularly to the involvement of lower management and relevant segments of the work-force, including representatives.

With the advent of participatory ergonomics, Professor Corlett's words are even more pertinent and in Chapter 3 we note that the role of the workforce, as well as management, is essential when collecting relevant information for a cost-benefit study.

We have already mentioned that Dr Paula Liukkonen of Stockholm University was instrumental in helping Swedish industry to examine the costs of occupational health and safety. We believe that she was the first to codify these costs in a form that enterprises could use to improve working conditions. Most of her work was published in Swedish but she has summarised some of her ideas in English (Kupi *et al.*, 1993).

Dr Arne Aarås has done some remarkable work in controlling and reducing musculoskeletal illnesses in a telephone wiring company, and has looked at the cost-benefit of this project (Spilling *et al.*, 1986). Comparing a seven-year period prior to the ergonomics intervention with a seven-year post-intervention period, he showed that there was a saving of about nine times the outlay for ergonomics improvements measuring only injury and labour turnover reduction. He did not calculate other productivity factors although some may have added to the cost-benefit ratio. The length of the study and its results, in both injury prevention and cost-effectiveness, have been a highlight in this field.

It is commonplace nowadays that before governments introduce safety legislation they make a cost-benefit prediction. This is often in response to industry saying that it cannot afford whatever is proposed – almost a knee-jerk reaction to any thought of expenditure to benefit workers. In a paper on the economics of the introduction of a safe manual handling code of practice (Oxenburgh and Guldberg, 1993) the methodology used was to determine what unsafe lifting was in terms of the code and thus what the code would do in preventing injury. A statistically determined cross-section of workplaces was surveyed by a team of investigators who measured the number of unsafe lifting practices (i.e. lifts addressed and 'corrected' by the code). The cost to correct each unsafe lift was determined and the cost expanded to an estimate of the entire industry and related to the estimated savings in back injuries. The limitation of the methodology is that small errors in the original measurements become greatly magnified in the expansion to an entire industry. As is so often the case, the authors had no opportunity to measure the actual costs after the code was introduced to determine whether the original methodology, and the estimates based on this methodology, were correct.

Professor Guy Ahonen has surveyed 340 small and medium-sized companies in Finland in order to examine and develop their occupational health and safety conditions. The results of the questionnaires were analysed through the use of the Productivity Model, and he compared the individual companies with the best in their industry group based on sickness absence; a 'best practice' methodology. Ahonen was then able to estimate the benefit to an enterprise of reducing the absence rate both in the actual savings and in productivity increase (Ahonen, 1998).

Professor Ahonen and his colleagues have also used the Finnish version of the Productivity Model to estimate the economic costs of alternative treatments of acute back pain, the costs of smoking and of ageing within the Finnish population. He has since developed a further computer model, derived from our earlier model, mainly for use in the Scandinavian market. It is known as 'The Potential' and is available in Swedish, Finnish and English (details are given in the References).

A very useful step forward in occupational health and safety costing was made by Dr Per Dahlén. He used activity-based costing to derive the cost drivers for injury and thus the allocation of those costs. Although he was looking mainly at the costs of absenteeism, labour turnover and training, he did introduce some productivity factors. These productivity factors were re-working, scrap and overtime due to the use of unskilled labour (Dahlén and Wernersson, 1995).

Dr Hal Hendrick, who has worked in industry for many years and is a past President of the International Ergonomics Association, has been very keen to see costing included in the evaluation of ergonomics interventions. He has collected some very useful case studies, his own as well as those of others, to illustrate this point and has published these cases through the Human Factors and Ergonomics Society (Hendrick, 1996). His message is that business needs ergonomics and costing is one method to drive the point home.

Dr Alison Heller-Ono has used cost-benefit analysis to substantiate the involvement of participatory ergonomics in the workplace (Heller-Ono, 2001). In several workplaces she has instituted an 'Ergonomics Task Force' encompassing up to a dozen workers from all levels from senior management to shop floor workers. One particular workplace was an office where the injuries were musculoskeletal to the upper limbs (cumulative trauma disorder). After an initial training period on basic ergonomics given by Heller-Ono, the 'Ergonomics Task Force' went to work to improve the conditions of the office workers and to meet safety and health standards. The method for financial analysis was return on investment (see Chapter 2) and Heller-Ono has followed this workplace over a four-year period. Each year of the process has been compared with the previous year and substantial benefits were observed each year, so that by the fourth year the return on investment was over $5 for each dollar spent.

It is unfortunate that although several people, as above, have published financial results from ergonomics interventions there has been very little development of generic models (models that can be used generally rather than specific to a particular situation). In 1996 Dr Marcel Knotter from NIA (The Netherlands Institute for the Working Environment) conducted a survey to see where and how many models are available, but the results of the survey have not been published.

A breakthrough to develop generic cost-benefit models that can be used by employers should have come about through a specific conference on the subject (Mossink and Licher, 1997). The papers at the conference were mainly regarding the costs and benefits one would like to measure in an enterprise/industry sector or on the effects of legislation. With the possible exception of two papers out of over 50, there were no general models to help *enterprises* implement occupational health and safety policies and procedures. The two papers were those of Dr Paula Liukkonen, who has a deep understanding of the factors that make up cost and benefits in industry (her model is not available in English), and the present authors' Productivity Model.

The fact that generic cost-benefit analysis tools are in short supply does not mean that some form of costing is not used by occupational health and safety practitioners, rather than only by research workers, as discussed above. For his thesis Michael Morrissey surveyed, by questionnaire, occupational health and safety practitioners in Australia (Morrissey, 2002). These people came mostly from larger companies and he found that about 75% of respondents had used some form of costing. However, on closer analysis the costing used was mostly based on the direct wage costs, with the indirect costs (lowered productivity, overtime, product damage and so on) often ignored. As we note throughout this book, it is often the indirect costs, not the direct costs, that are major cost factors and it is from savings in these indirect costs that benefits flow from an intervention.

Morrissey's survey, valuable in itself, points to the unfortunate fact that cost-benefit analysis is not generally used by occupational health and safety practitioners and, with the possible exception of Finland, we think that this may be the situation generally.

Although we suspect that cost-benefit analysis models for occupational health and safety have been developed by some large enterprises for their internal use, they have not been published or otherwise made generally available. It seems unfortunate that there are not more models developed for occupational health and safety on the market so that users would be able to choose that which best fits their needs. Models that are *not* suitable are, for example, engineering ones that derive only from the technology of the design or process and do not express the effect on, and the effect of, the workers and other persons concerned. It is at least partly to fill this gap that the Productivity Assessment Tool (and its predecessor, the Productivity Model) has been developed.

1.4 Productivity Assessment Tool

In a review of the first edition of this book a reviewer (Thompson, 1995) commented that we had overlooked the area of 'senior management commitment to safety as a business objective equatable to quality and profit'. If senior management really has this equal commitment to safety (a rare quality in practice although not in word) then cost-benefit analysis is not required. Cost-benefit analysis for occupational health and safety practice is a tool to convince otherwise sceptical management that the value of *people* must not be underestimated and that workers are their key to profit.

The emphasis in the Productivity Assessment Tool is on employees and the costs and benefits that their employment brings to the enterprise. It is not the only tool, but one of many, that can be used to bring home the notion that people should be treated humanely and work safely. It is reasonably obvious that the needs of users will vary from industry to industry and no claim is made that the Productivity Assessment Tool is complete or will suit all needs. It is up to the user to use the tool intelligently and adapt it to his or her needs.

In the first edition there were 60 case studies illustrating occupational health and safety solutions to poor working conditions accompanied by a paper version of the Productivity Model. A paper model is difficult to use, with too much arithmetic for the user, although this was later overcome by making the model into a computer program. The software for this program has been through three versions and a basic version of the latest one, productAbilityBasic, is enclosed with this edition.

The simplest and most frequent use of the Productivity Assessment Tool is to measure the changes in the effectiveness of employees in an intervention. That is, it is time-based and can compare a 'before' and 'after' situation.

The model can also be used for analysis of, and comparison between, two or more proposed interventions; in this case it can assess the cost-effectiveness for each intervention although it will not determine which is the most effective in injury prevention terms; that is not, of course, the purpose of the model.

An interesting feature of the model is the use of sensitivity analysis. It is often not realised which are the most sensitive cost parameters and, by entering the workplace data, variations in this data will demonstrate the role of each cost parameter (overtime, down-time, injury absence, etc.).

As a bonus, the cost-effectiveness of rehabilitation for even a single worker can be estimated by the use of this software and its use is illustrated in a case study in Chapter 5G.

Use of the Productivity Assessment Tool is more fully described in Chapter 4; Chapter 5 gives case studies to illustrate the use of the program. Where you get the information for your *own* analysis is described in Chapter 3, and the basic economics that will assist you to follow the reasoning is in Chapter 2.

1.5 References

Ahonen, G., 1998. The nation-wide programme for health & safety in SMEs in Finland. Economic evaluation and incentives for the company management. *In From Protection to Promotion. Occupational Health and Safety in Small-scale Enterprises. People and Work.* Research Report 25, Finnish Institute of Occupational Health, Helsinki, pp. 151–156.

Corlett, E.N., 1976. Costs and benefits from human resources studies. *International Journal of Production Research*, **14**, 135–144.

Corlett, E.N., 1988. Cost-benefit analysis of ergonomic and work design changes. In Obourne, D. (ed.), *International Review of Ergonomics* (London: Taylor & Francis), vol. 2, pp. 85–104.

Dahlén, P.G. and Wernersson, S., 1995. Human factors in the economic control of industry. *International Journal of Industrial Ergonomics*, **15**, 215–221.

Health and Safety at Work: a Question of Costs and Benefits, 1999. Office for Official Publications of the European Communities, Luxembourg.

Heller-Ono, A., 2001. Successful outcomes of an ergonomics process using an ergonomics task force. *Proceedings of the 37th Annual Conference of the Ergonomics Society of Australia*, pp. 129–136.

Hendrick, H.W., 1996. *Good Ergonomics is Good Economics* (Santa Monica, CA: Human Factors and Ergonomics Society).

Kupi, E., Liukkonen, P., and Mattila, M., 1993. Staff use of time and company productivity. *Nordisk Ergonomi*, **4**, 9–11.

Morrissey, M., 2002. *Assessment of the extent that OHS professionals use financial analysis to evaluate OHS programs.* Masters thesis, School of Safety Science (Sydney: University of New South Wales).

Mossink, J. and Licher F., 1997. *Proceedings of the European Conference on Costs and Benefits of Occupational Safety and Health* (Amsterdam: NIA TNO).

Oxenburgh, M.S., 1991. *Increasing Productivity and Profit through Health and Safety* (Sydney and Chicago: CCH).

Oxenburgh, M.S., and Guldberg, H.H., 1993. The economic and health effects on introducing a safe manual handling code of practice. *International Journal of Industrial Ergonomics*, **12**, 241–253.

Pyle, J., 2002. Keynote address to the *Women Work & Health Conference*, Stockholm.

Spilling, S., Eitrheim, J. and Aarås, A., 1986. Cost-benefit analyses of work environment investment at STK's plant at Kongsvinger. In Corlett, E.N., Wilson, J.R. and Manenica, I. (eds), *The Ergonomics of Working Postures* (London: Taylor & Francis), pp. 380–397.

The Potential. Cost-benefit analysis software available from Miljödata at info@miljodata.se or www.miljodata.se.

Thompson, K., 1995. *Ergonomics*, **38**, 421–422.

2 Economics for the non-economist

Contents

2.1 What this chapter is about

This chapter provides an introduction to the language and ideas of economics and gives a brief sketch of the concepts used in economic analysis; it also shows how these principles have been incorporated as a foundation of today's accounting and financial control systems. It introduces the Productivity Assessment Tool as an economic analysis model, a tool that can bridge the gap between human resources, occupational health and safety and financial systems.

With these economic concepts under your belt you have the fundamental framework to take an active part in debates in your enterprise about the relative merit of spending proposals and, in particular, to analyse critically the cost impact of decisions that include, or even omit, considerations of workers' health, safety and welfare.

Fundamental to these debates is an understanding of the way assumptions and generalisations are built into economic models. These assumptions may not always be stated in an economic analysis but they are the basis of the economic theory that underlies all models, including financial control systems. This chapter explains these assumptions and what they mean to the 'bottom line' as well as to occupational health and safety and human resources management. It also makes reference to the parts of an economic model that are included in financial control systems and accounting models.

It is generally assumed that projects provide returns over a relatively short term. At the end of the chapter a section is included that considers the impact of long-term projects. This necessitates calculating how the value of money will change over time, and includes more mathematics.

2.2 Why economics?

A dictionary definition of economics is 'the science of the production, distribution and use of income and wealth'. Simplistically economics is the study of money, and money is often the reason given for funding or not funding a particular project. By understanding the economics of investment decisions on one side and the business accounting systems built to track enterprise financial performance on the other, it is possible to understand the financial decision-making process. Moreover it is possible to present projects in terms of their cost advantages to the enterprise and to analyse other projects in the same way.

To focus on financial decision-making does not suggest you should ignore other reasons for projects being accepted or rejected, such as internal or external politics and the relative power and influence of the proposers and champions of the project. Rather we suggest you use financial analysis as another tool to assist in asserting your point of view.

The economist Paul Samuelson (1980) provided the textbook definition of economics as:

> the study of how people and society end up choosing, with or without the use of money, to employ scarce productive resources that could have alternative uses – to produce various commodities and distribute them for consumption, now or in the future, among various persons and groups in society. Economics analyses the costs and the benefits of improving patterns of resource use.

On an economic basis a project is worth funding when its value to the enterprise is greater than its cost. Where several projects are competing to be implemented the project that provides the greatest value is considered to be the most worthwhile. At first glance the concept is simple; in reality it is anything but. In reality it is possible to argue about every item of cost and every prediction of benefit that is included in a project proposal. Large-scale projects that run for long periods are the most difficult to cost as they need to consider the impact of factors outside the enterprise, such as changes in interest rates. Luckily most ergonomics interventions have short pay-back periods, making them easier to cost.

This chapter explores one commonly used source of cost data, the enterprise's accounting and financial management systems. Accounting systems are based on economic concepts but are narrow in focus, constructed for the purposes of taxation, regulation and financial accounting. In this book we use the term *economic model* as a generic term to include accounting and financial models. Traditional accounting practice considers workers only as an expense; this chapter explores the implications of this for any project that invests in workers. Chapter 3 explores these data collection issues, focusing on determining what data are relevant for inclusion in your analysis.

Benefit data is based on prediction of expected outcomes from the investment. These may need to incorporate predictions derived from economic modelling of the future of the economy, whether at a local level or globally.

2.3   Predicting the future

Economic predictions, forecasts, are used to help enterprises plan for their future. For instance, predictions of economic growth and interest rates may be used to predict an enterprise's sales and hence determine its production targets. You may well ask: how can economists take into account everything that happens in an economy to make accurate predictions? Of course, this is not possible; in the real world many factors vary together and economists can only make guesses at the impact of any one factor. Economic theory makes guesses or generalisations, termed *assumptions*, about how groups of workers, industries or economies will behave. These assumptions are in turn incorporated into computer-modelled economic 'experiments', the results of which are necessarily imprecise; hence economics is an *inexact* science.

Economic predictions are built on the experience of what has happened in the past and how the economy has reacted to change, e.g. interest rates rising leading to slowing of the housing market and reduced demand for consumer durables such as stoves and refrigerators. In this way, economists use the past to predict the future.

Where events have no precedent economists, like everyone else, have no basis for making predictions. For example, the development path of computers was not foreseen by many central figures in the fledgling industry. Back in 1943 Thomas Watson, Chairman of IBM, then one of the world's leading adding-machine companies, predicted: 'I think there is a world market for maybe five computers' (Farrell, 2001). In 1977 Ken Olsen, President, Chairman and Founder of Digital Equipment Corp. predicted: 'There is no reason anyone would want a computer in their home'. Bill Gates, Founder and CEO of Microsoft, commenting on computer memory size predicted in 1981: '640K ought to be enough for anybody'. Just think of the range of predictions that surrounded 'Y2K'. With so many computer systems that had been developed with 2 digit year date fields no-one knew what impact the change of date

Insurance prediction – hazardous guesswork

The financial viability of the insurance industry is based on prediction and quantification of risk. In the 1930s Lloyds of London, an insurance institution with a 300-year history, began writing insurance policies for asbestos producers. The policies did not have any limitations or restrictions regarding asbestos. In the same decade two major workforce-based studies carried out among textile workers, one in the United Kingdom and one in the United States, provided evidence of an exposure-response (and therefore likely causal) relationship between level and duration of exposure to asbestos and radiographic changes in the lungs indicative of asbestosis. These reports formed the basis of the first asbestos control regulations in the United Kingdom, promulgated in 1930, and the first threshold limit values for asbestos published by the American Conference of Government and Industrial Hygienists in 1938 (Selikoff and Lee, 1978). Thus at the time Lloyds began insuring asbestos producers there was information becoming available that asbestos posed a health risk. Despite the increasing availability of such information Lloyds continued to insure asbestos producers over the subsequent decades.

The first damages case from work with asbestos was won in 1969 and by the year 2000 payouts for such claims had totalled billions of dollars; Lloyds itself was severely financially drained by payouts on asbestos claims and thousands of their investors (the 'Names') were ruined:

> Lloyd's of London had written liability insurance for American asbestos companies since the 1930s [actually, reinsurance on their insurers], . . . the policies were still in effect, and . . . were generally unlimited. There were no maximum amounts where the insurance stopped paying, and no diseases were excluded from coverage. The potential damage awards – and Lloyd's exposure [it had insufficient reinsurance itself] – were limitless, (McClintick, 2000).

from 1999 to 2000 (from 99 to 00) would have. Predictions ranged from there being no impact to there being total economic chaos.

Economists also develop predictions on the basis of different schools of economic thought. Each school of economic thought makes different assumptions about how the world of money functions. An enterprise seeking to use economic predictions to assist in business planning will find that different economists, and different schools of economic theory, will give quite contradictory views on the economic outlook. Economists are working with only a small part of the economic jigsaw puzzle, and need to make assumptions about the rest. If they can only identify factors that affect 15% of wage movements observed in the real world (as is the case with some economic models) then the rest is a hazardous guess. Just how hazardous the guesswork can be is seen each year in spectacular corporate collapses.

If you are thinking 'I never have to deal with economic predictions', you may be wrong. As there is no certainty about the future we must rely on predictions. Even small-scale projects that are completed in a short period need to develop specific predictions about the outcome of that project.

On a larger scale, economics is the basis of every enterprise system used to measure and track money, workers and products including accounting, budgeting and financial systems. Economic models as assessment tools are increasingly being used to evaluate the worth of alternative proposals for large-scale enterprise or government spending. Understanding what is included and what is left out of these models will help you to argue on cost terms for your point of view and for your proposals.

2.4 What are economic models?

Economic models are *tools* used to test economic theory. They apply the theory to particular situations, usually to predict what will happen in the future. The models represent the economic world, or a part of that world, but are simplified derivatives. They are mathematically based and work by translating economic factors into numerical values so they can be added, subtracted, multiplied or divided together to give a numerical answer. The numerical values chosen:

- *simplify the calculations* required within the model by assuming workers or enterprises will *always* behave in a particular way, e.g. it might be assumed that if the enterprise raises its prices by 5% it will lose 10% of its sales volume
- *quantify abstract concepts* to enable these concepts to be included in calculations, e.g. placing a monetary value on the goodwill of an enterprise.

As the models are mathematically based they easily incorporate measurable data such as wages but have difficulty incorporating abstract concepts such as goodwill or worker motivation. It is easy to measure the wages paid to workers, but measuring the value gained in return is more complicated. In neo-classical microeconomics, differences in wages are assumed to reflect totally differences in quality and value of each worker. This assumption has been built into accounting systems where workers are considered only a cost to the enterprise. In subsequent chapters you will see that to gain an approximation of the value of worker productivity you need to measure a much broader range of factors than just wages. The Productivity Assessment Tool has been specifically designed to capture the value workers add to the enterprise.

The economic world is deemed to be made up of four sets of factors or *resources* contributing to the production, distribution and use of income and wealth:

- human resources – labour
- natural resources of the earth including minerals, trees, water and air
- money resources – income, capital and wealth
- products and services produced by enterprises.

Economists develop theories to examine and explain the relationship of these factors, their variation over time, how they will react in different situations and their interaction in the marketplace.

The study of these factors within individual enterprises or individual industries is termed microeconomics. The following are some examples of situations where microeconomics is used to examine:

- the way the price of a product changes when you change its availability. To take an example from everyday life, if vegetable crops are destroyed by rain then the prices of vegetables will rise as they are in short supply. If more farmers plant one type of vegetable and they all arrive at the market together then the price will fall as farmers compete to sell their vegetables
- the costs of products or services using different production methods including varying the ratios of labour and equipment. For example, the use of an automatic darkening welding helmet made the task of spot welding at least one third faster for a bed manufacturer (see section 5F.3, page 124). This enabled the manufacturer to stay cost-competitive against imports.

On a larger scale economists theorise about the economies of entire countries or the world economy, termed macroeconomics, and seek to explain the interaction of all the resources within the economy and to forecast their future direction. Some situations where macroeconomics is used are:

- forecasting economic growth (or recession) in national and global economies
- making individual enterprise-based predictions of future sales and expenses including such factors as growth or reduction in its target market, changes in cost of enterprise borrowings, changes in costs of imports/exports due to exchange rate variation
- driving national governments' decisions about fiscal policy – control of spending through changes to taxes and government spending
- driving national banks' monetary policy – control of the money supply through changes to interest rates
- examining the impact of globalisation on trade and poverty.

2.4.1 Microeconomic modelling

Microeconomic models are used by individual enterprises or industries to analyse the relationship between the production and supply of products, the demand for those products and their market price. *Products* in this context are anything sold by the enterprise, whether physical items or services. An understanding of microeconomic modelling in general will elucidate how production decisions are made and what factors enterprises consider when they look at their production costs.

Microeconomic models are also used to analyse the enterprise or industry's market for other economic factors: human resources, natural resources, income and expenditure.

The Productivity Assessment Tool is an example of a microeconomic model. It is designed to measure in detail the contribution of workers to the production process, and its method of use is described in Chapter 4.

The components of microeconomic models can be considered in terms of:

- production and supply of products or services
- demand for products or services
- the market price.

Production and supply of products and services

It is assumed that enterprises will produce their products or supply their services using the cheapest and most efficient method available so that they can supply their products to market at the lowest profitable price. All enterprises are assumed to behave similarly so that they compete with each other on an equal footing to supply products and services to the market. In some situations consideration may be given to standards of production, but enterprises are assumed to use the cheapest production method to achieve that particular standard.

Production costs are the cost of labour (wages and indirect costs) and capital (equipment, technology, cost of raw materials, etc.) required to produce and supply the product or service. Measured costs may leave out the costs of developing worker skill levels or costs of faulty products and waste.

The assumption made by economists that enterprises will use the cheapest and most efficient methods available to produce their products and services does not always ring true. Often enterprises continue using old and less efficient machinery or systems and they may choose, for a variety of reasons, not to replace old equipment or systems. For instance, if the economic assumption that enterprises will use the cheapest and most efficient methods available had been true, then the most efficient means of patient handling would have been put in place in health facilities as the technology became available. Instead, it was not until the cost of injuries to nurses and shortages of nursing staff was factored into service costs that health facilities began to install sufficient quantities of lifting devices and patient handling systems (see the health industry case study, Chapter 5C).

Ergonomists and human resources managers are often called upon to assist in making older production systems safer and more productive through changes to the workers themselves, such as changing work practices or training. For instance, it is seen as cheaper to train workers to lift than to purchase lifting equipment. However, when all production costs are considered, such as the cost of reworking faulty products and costs of injuries, then it can be seen that in fact it is often more cost-effective to change the production system.

Prevention need not be expensive

The buying clerk in a company is often given the responsibility of reducing costs (spending as little as possible) but this may be at variance with the safety needs. In a common law case in Australia in which one of the authors was involved, a serious back injury was caused by a worn-out hand tool (a plumber's wrench) slipping and giving way. Although a new tool had been requested over the years it had been turned down by the buying clerk on the grounds of expense – after all, there already was a plumber's wrench in the workshop. The outcome of saving perhaps $100 on a new tool was a payment of a quarter of a million dollars to the plumber for his injuries: a high cost to the employer and insurer and an even greater cost to the worker as no amount of money is able to make up for his pain-ridden life.

In the simplest microeconomic models it is assumed that enterprises can switch at any time, without extra cost, between the use of labour, machinery or technology and they can adjust the total hours of employment they offer on a daily basis. The important thing about this assumption is that under economic theory it is an enterprise's responsibility to minimise production costs but there is no mention of any reciprocal responsibility to provide workers with a living wage. For example, increasingly service industries such as call centres and hotels use casual workers who can be rostered to work on relatively short notice and/or for short periods of time. This is not necessarily the best situation for the workers involved, and is considered in more detail in Chapter 5.1.

In accounting systems production costs are often considered simply in terms of the ratio of labour costs compared with capital costs with the focus on minimising costs, thus favouring changes (projects) that directly reduce these costs. Such analyses can leave out significant factors in the cost of production.

Projects directed at other factors such as reducing injury risk, improving management systems and reducing waste may achieve reduced costs of production through improved productivity. Under accounting models the total costs and benefits of these projects may each be accounted for differently, preventing a true comparison of the total costs versus total benefits. For instance, some production costs are commonly accounted for as enterprise-wide costs or overheads. These may include:

- warranty or product defect costs
- workers compensation costs
- absenteeism costs
- turnover costs.

By contrast, expenditure aimed at reducing these costs may be accounted for against the local product or service area. The perceived impact of a local intervention can be diluted if the improvement is accounted for as an enterprise-wide improvement. Example 2.1 illustrates how the method of accounting for costs and benefits can change the *perceived value* of a project.

Demand for products and services

Demand has a special meaning in economic theory and it is used to equate to *buyers* of a product or service. It is assumed that when prices are high very few people will be prepared to buy products. As prices fall more people will be prepared to buy, e.g. in the 1980s when personal computers were first developed they were scarce, prices were high and very few people bought them, but at the beginning of the twenty-first century their price has fallen dramatically and personal computers are ubiquitous.

Underlying the theory of demand is an assumption that if an enterprise wants to sell more products or services then it is necessary to reduce production costs and hence prices. The argument that investing in workers adds to production costs and pushes up the price of products and services is often given as the reason that training, ergonomics interventions or equipment replacements are not funded. As mentioned in the previous section, when you consider their total impact there may be a net gain from these projects even though some costs increase.

Example 2.1: The perceived value of a project

This example illustrates a common problem with the calculation of the cost-benefit of a project, comparing apples with oranges. Following a quality audit it was determined that 50% of all product defects at Hearthstone Ltd occur on one of ten production lines. Therefore the quality manager wants to allocate funds for a quality assurance program to reduce product defects on this one line. The quality manager has to justify the project to the line manager because the funds for the program will come from the line manager's budget.

They both agree that as a result of the program the production faults on that one line could be halved; this represents a plant-wide reduction in defects of 25%. Given that there are ten lines this is also an average 2.5% reduction in defects per line.

The line manager says that the project is too expensive as the benefits to the one line are only a 2.5% reduction in defect costs but the quality assurance program will cost more than that to run. The project does not seem to be cost-effective.

The problem is one of accounting. Within Hearthstone Ltd's accounting system the estimated reduction in defects of 25% may be recorded in either of two ways:

- if Hearthstone Ltd accounts for product defects as part of general overheads the 25% reduction in defects at the plant will translate into an average of 2.5% reduction in defect costs per line, or
- if Hearthstone Ltd accounts for product defects against each production line then the full 50% reduction in defects will be recorded against that one line.

The accounting method chosen is immaterial if funding for the quality assurance program is accounted for in the same way:

- where 10% of the improvement in quality is allocated to each of the ten lines' budget, only 10% of the quality assurance program cost should be allocated to each line's budget, or alternatively
- where the full 50% improvement in quality is allocated to that local line, the total cost of the quality assurance program (i.e. 100%) should be allocated to that line's budget.

The problem arises in this case because the line manager has compared the benefit of the project spread over the entire plant (2.5% improvement per line) with the total cost of the program allocated to one line's budget. The total benefit to the enterprise has not changed, but from the perspective of one line's budget the project looks expensive and will not be funded.

When you are calculating the pay-back period of a project it is important to calculate costs and benefits so they are directly comparable, i.e. both calculated locally or both calculated enterprise-wide. That is, compare apples with apples, not with oranges.

Selling price of products and services

The price an enterprise will receive for its products is considered to be determined by bringing together buyers and sellers in a market. The enterprise will offer its products for sale, and if the products match buyers' expectations of price they will be sold. In economic terms, when the supply of a product equals the demand for that product, this will determine the selling price. This is termed the *market price* or simply the *price*. The product price and each enterprise's market share also determine the quantity that an industry will produce.

Similarly, workers are considered to supply their labour to the marketplace and take the wages offered. Economic news reports will refer to this relationship between supply and demand as 'economic forces at work' or the 'market at work'.

In microeconomic theory it is assumed that, under perfect competition, enterprises cannot affect the market price by increasing their output or differentiating their product. This is because there are assumed to be no barriers that stop others entering the same market and competing with existing enterprises. Economists find it difficult to point to real-life examples of perfect competition.

For most enterprises there is some extent to which their products are differentiated from their competitors, and then it can be possible to make more than the bare minimum return. For instance, vehicle manufacturers in Sweden during the past several decades found that improving the skills of their workforce and giving workers the responsibility for quality resulted in improved quality of cars manufactured. Not only has this saved warranty costs but buyers have been willing to pay more for a higher quality product and there is more demand for these higher quality cars.

When arguing for funds, whether to improve workers' safety, motivation or conditions, consider the likely impact of the changes on product sales. Where you cannot argue that your proposed intervention will make the final product cheaper to produce, you may be able to argue that buyers may be prepared to pay more for your product or service depending on its quality.

Limitation – social costs not included

Microeconomic models consider costs borne by the enterprise and are limited to consideration of the effects of manufacture or sale of a product or service. Because the total impact of workers' labour is not incorporated in microeconomic models, the full value of human endeavour can be missed. For example, if an experienced nurse suffers a severe back injury, is forced to leave nursing and is permanently restricted in her personal life, microeconomic modelling does not cost the:

- productivity loss from the loss of that nurse's experience, including relationships with patients and nursing skill
- social costs of that nurse's pain and suffering and restrictions on activities outside of work.

These latter factors are *social costs* in economics.

The Productivity Assessment Tool is a microeconomic model and so does not take into account the effect of social and other outside costs. Although this is a limitation to the model, as the assumptions made within the model are relatively few it means that the model can be more easily and simply used.

There will be situations where the social costs are significant, including occupational diseases. Due to the long lag time between exposure and the development of occupational diseases, such as asbestosis, silicosis or hearing loss, it can be difficult to prove that the employer was responsible and thus the cost is borne by individuals and their society. From the enterprise's perspective, if workers are at risk of occupational disease in 15 or 20 years' time, the enterprise may not include an allowance for the costs of future treatment in the current costs of production. Compared with the potential short-term gains, the long-term risks may seem too far off to worry about. This has two effects:

- it makes it more difficult to argue in support of expenditure to prevent exposure to the hazardous substance within the workplace
- it means that the costs of diseases that occur will be borne by the worker, the family and social services, i.e. the whole community.

In these situations where there is a need to include costs that are outside the enterprise or are long-term you will need to use another tool, macroeconomic models, for your analysis (see sections 2.4.2 and 2.6.2).

Accounting programs are often used as decision-making tools within organisations, and it is likely that other people will use them to prepare proposals. Accounting programs, although based on microeconomic models, are often focused on taxation requirements and not on other important parameters of production or service.

In any situation where your proposal is competing for funds against other projects it is important to prepare your data on the same basis as those other projects, or you could argue that the other projects should calculate their costs on the same basis as yours.

2.4.2 Macroeconomic modelling

Macroeconomic models attempt to explain the supply and demand for products across an entire economy: regional, national or global. The models include assumptions about factors that affect a whole economy, including:

- changes in interest rates
- savings patterns
- investment patterns.

Macroeconomic models seek to capture the total effect of the changes they consider. For example, macroeconomic models include changes that are not usually funded by industry, including:

- pollution and environmental effects
- costs of chronic work-related illness and disease
- costs of unemployment
- costs of training and retraining
- other social costs.

Varying macroeconomic assumptions leads to different results

This insert looks at interpretation of macroeconomic modelling undertaken to examine the impact on the Australian economy of removing tariffs from the vehicle manufacturing industry.

In the mid-1970s it was debated whether or not the 30% Australian import tariff for motor vehicles should be removed. The government of the day opted for a conservative, gradual, stepped tariff reduction program to a 15% import tariff by 2000.

Looking back over the changes in Australian vehicle manufacturing between the 1970s and 2000 the Department of Foreign Affairs and Trade (2000) stated that individual vehicle manufacturing firms have responded to cuts in tariffs by:

- increasing productivity
- making significant improvements in quality, and
- expanding exports significantly with many Australian manufacturers becoming internationally competitive for the first time.

Looking at the same economic data an Australian economic think-tank, NEIR (2000), came to the opposite conclusion. Their interpretation is that:

- the costs to the Australian economy of reductions in tariffs are greater than the benefits. These costs in terms of reduced levels of employment and economic activity have been disproportionately borne by manufacturing industry and manufacturing workers
- for Australia as a whole, tariff reductions have increased the inequality of income distribution
- tariff reductions have made the economy more not less dependent on mining, natural resources and primary industry and
- furthermore, the reduction of tariffs by 2000 has resulted in the loss of approximately 60,000 manufacturing jobs.

The data available for analysis was the same for both, and both used the same macroeconomic model (Orani) to calculate the impact of the tariff reduction. The difference between the groups was in the assumptions that each analysis made and this led to them making completely opposite conclusions about the impact of tariff reductions. As one economist puts it: 'significantly different results . . . can be obtained without any change to the structure of the model. It suffices to change the parameters (assumptions) which is why the Orani model can produce different answers to the same question', (Manning, 1998).

Governments make use of macroeconomic models in their decision-making process as they have responsibilities that cover society as a whole. It is common for governments to use models to study the cost effect of new laws, regulations or standards before they are introduced.

Macroeconomic models are a mathematical rendering of an economic system. Where information or variation (the actual data) is left out of the model, being too difficult to collect or just not available, economists replace that data with assumptions.

Due to the scientific, mathematical or accounting nature of the calculations, at times these models are presented as providing the unambiguous answers to the questions posed. But while the calculations are mechanical processes, it is the choice of assumptions and data values that go into any model that reflects the economist's point of view and these will bias the outcome. Because macroeconomic models are not value-free, you may disagree with the way a model has been constructed including any or all of its assumptions, and hence you will also disagree with answers produced from the model. This applies to all models whether describing economics, injury causation or bridge-building. As computer programmers succinctly put it, 'rubbish in, rubbish out!'.

On examining the cost effect of a new law or regulation it is important to consider the assumptions made, as the answers macroeconomic models provide are limited by the assumptions built into them. If the assumptions are not reasonable then the results from the analysis will be flawed. As the models grow the assumptions become more numerous and complex, making it harder to see the limitations in the model.

The assumptions made and the value of the variables should be specified, and are often listed as an appendix. The phrase *ceteris paribus* is economic shorthand for 'all else remaining the same' indicating that the model is based on the standard assumptions (see section 2.5).

Solely on the basis of its assumptions, the answer provided by a macroeconomic model can vary from support for a project to rejection of a project. The next time you look at an economic or financial report, or are part of constructing one, take a critical look at the assumptions it is built from, not just the bottom line. If the assumptions do not make sense then the answer the model provides may be totally irrelevant. You can have a specific numerical answer, but not be sure what the number is measuring. After all, as Douglas Adams (1979) wrote in his book *The Hitch Hiker's Guide to the Galaxy*:

> . . . a race of hyper intelligent pan-dimensional beings once built themselves a gigantic supercomputer called Deep Thought to calculate once and for all the Answer to the Ultimate Question of Life, the Universe and Everything . . . For seven and a half million years Deep Thought computed and calculated, and in the end announced that the answer was in fact forty-two and so another, even bigger computer, had to be built to find out what the actual question was.

2.5 The assumptions and limitations of economic models

2.5.1 How reliable are economic models?

It is not possible to build an economic model that is an exact replica of the real world. The real world is just too complex for that to be feasible. Imagine the time it would take to find out how each person in the enterprise felt about every issue you might want to investigate, even assuming they were willing to tell you, or asking customers to tell you how much they are really prepared to pay for your products. Instead it is necessary to include some generalisations that will make the economic analysis feasible.

The critical question is not 'do I need assumptions?' but rather 'what assumptions do I make?'.

As noted in section 2.4, when gathering information generalisations will need to be made to quantify economic factors. The numerical values chosen:

- *simplify the calculations* required within the analysis by assuming that workers or enterprises will *always* behave in a particular way
- *quantify abstract concepts* to enable them to be included in calculations.

The values chosen must make sense within the problem under consideration. The analysis should take account of all the factors you think will affect the outcome. Where no data is available, either because the issue is too complex or data is missing, you will need to make an estimate for your analysis.

One way to reduce the error in your analysis is to make fewer sweeping assumptions. For instance, in the Productivity Assessment Tool worker productivity is calculated on the basis of a number of measurable sub-components and the data only considers the affected work area or work station within an enterprise, thus reducing the number of factors to be considered. Another way to improve accuracy is to quantify more factors, incorporating the actual data rather than estimates. The Productivity Assessment Tool is based on reducing assumptions and increasing the verifiable data.

The Productivity Assessment Tool is also designed to show how workers contribute to your enterprise's productivity; after all, without people there are no profits. If you look at your enterprise's annual report you will see items listed as either costs (expenses) or assets on the balance sheet and profit and loss account. The items that measure workers' contribution to the enterprise often will only appear on the expense side of the ledger under wage costs.

Increasingly, service-based enterprises are incorporating the skills and experience of their workers as an asset, i.e. incorporating the value that a skilled workforce brings to the enterprise on the asset side of the ledger. If workers are seen as only a cost to your enterprise then it is difficult to put forward any argument for extra expenditure on worker occupational health and safety, training or welfare.

You will still find factors that cannot be quantified, e.g. worker morale, and they need to be acknowledged and incorporated into your project submission.

2.5.2 Key general assumptions of economic models

This section could go by the alternative title 'how to avoid having the wool pulled over your eyes'. This is because, alongside the assumptions made consciously as part of an economic analysis, there are many assumptions underlying economic theory that are considered to apply across all economic and financial systems. These are also included as part of the model, either to simplify it or as estimates of difficult-to-cost factors. However, these assumptions may not be specified in any documentation and the phrase *ceteris paribus* indicates that, besides any specified assumptions, the general assumptions of economics are also considered to hold.

Economic theory argues that in a perfect world these assumptions are all true, but in the real world it is a matter of debate whether they are true and, if so, in what specific context.

When an enterprise evaluates the advantage of undertaking a new project it will do so using accounting systems, financial systems or marketing systems that have been built on these same assumptions, and there is rarely any debate about the validity of the system. It is precisely because these systems have been built on sweeping assumptions (including 'there are always more workers' so, by implication, you do not need to take care of what you have) that concerns raised on occupational health and safety or human resources grounds may not be properly considered. Unlike accounting-based models, the Productivity Assessment Tool has been formulated to make the value of workers and their contribution to the enterprise's productivity explicit.

What if the assumptions are not true? Where the general assumptions of economics do not hold true, economists consider the market to have 'failed' and hence these situations are referred to as *market failure*. In practice, this occurs due to a variety of factors. Market failure may result from:

- *lack of information* or uneven knowledge between buyers and sellers of products or services
- *lack of competition*, where there is only one enterprise, or a small number of enterprises, that provide a particular product so they can charge any price for their product and buyers have no choice but to pay that price if they want the product
- *price distortions*, where tariffs or other taxes make imported goods more expensive relative to local products or the production costs of local products are subsidised so they can be sold more cheaply
- *externalities*, situations where enterprises do not pay the full costs of their production, e.g. pumping waste into local rivers but not paying to clean the water or dismissing workers who report injuries to avoid paying workers' compensation costs
- *public goods*, factors that provide benefits to all enterprises which no one enterprise wants to pay for, e.g. provision and maintenance of roads
- *capital market imperfections*, flaws in the market for money, e.g. difficulty obtaining loans to fund purchase of new or more efficient production equipment and systems.

A list of general assumptions follows, and subsequent sections provide lists of assumptions that are specific to workers, producers and consumers.

- *There is 'perfect' knowledge for consumers, sellers and workers*, meaning that no-one has an advantage over the other and no-one is acting on poor information. Whether this assumption is true is often a critical issue for occupational health and safety. We have often met workers with back injuries who say 'if only I had known what it would be like to have a back injury I would have approached work differently', implying that they had not really understood the consequences of injury; they were acting on poor information.
- *Workers make fully informed decisions about their employment conditions and will freely move to another employer if not satisfied.* This assumes workers are fully informed about things such as the risks in their jobs and wage rates available elsewhere and that, individually, they have the power to bargain for the pay and conditions that they deem acceptable. It also assumes that they are free

to move from place to place in search of work. If the assumption is true then there is no need to provide additional work conditions or remuneration above those currently accepted by workers, nor are collective worker organisations (trade unions) necessary. Trade unions argue to the contrary; that individuals do not have the power to bargain for better pay and conditions.

- *There are many small suppliers competing to sell similar products where no one enterprise can influence the price they get for their products.* Small butcher shops or greengrocers in a local shopping centre are often given as examples of this type of supplier. If this assumption is true then enterprises could not raise their prices for products without losing sales to a competitor. In today's world with globalisation of enterprises it is increasingly the case that there are only a small number of suppliers and purchasers of products. In some cases suppliers (as monopolies) have the power to set the price of their products to consumers or to significantly influence their prices. In other cases it is the large chains of retailers that have the power to dictate the market price, e.g. Wal-Mart in the USA.

- *All the enterprise's costs in producing a product or service are included in the price it offers to the market.* As well as the cost of raw materials and the production process the price reflects the other factors, often grouped together as overheads, that are part of the cost of production – warranty costs for the product, workers' compensation costs, workers' entitlements, waste disposal, etc. This can be a problematic assumption as sometimes part of the costs of producing a product is borne by society as a whole. For instance, if an enterprise pollutes the air leading to acid rain, neither the effects of the acid rain nor the costs of its clean-up are included in the production costs. Another common example is the costs of occupational diseases, many of which take decades to become apparent by which time a link cannot be proven between the enterprise, where the disease was contracted, and the disease: the costs of treating the illness thus falls on the society.

- *There is no scarcity of resources* or, for resources that are less available, *their price is considered to reflect their scarcity.* This assumption came about because current economic theory was developed in the 1930s when the world still seemed a big place with abundant resources. Following this logic there are always more workers, more trees, more minerals and more clean water. Today the assumption is questioned on all sides; think of the whale and fish species hunted to the point of extinction, or the degradation of the Black Forest in Bavaria and other forests in Europe from acid rain. This assumption is of particular importance to human resource and occupational health and safety practitioners because, by implication, workers are just another resource in abundant supply. There are more children being born every minute and if workers are no longer useful they can simply be discarded and replaced. Under this assumption there is no necessity for enterprises to train and develop their workers or prevent work-related injury.

- *Today is more important than tomorrow.* It is assumed that an enterprise would prefer to use all available resources this year rather than allocate funds to keep them for future years. This assumption was developed during the post-First World War days and the depression of the 1930s, when tomorrow seemed uncertain. Today the assumption is echoed by the short-term tenure of many chief executive officers and senior managers and by the daily pressure on stock prices. There is an increasing trend for senior managers to achieve short-term goals and then move to another organisation, mitigating against the funding of long-term projects requiring a long pay-back period. In many industries it is becoming increasingly difficult to gain approval for projects perceived to be long-term, although case studies show that ergonomics projects usually pay for themselves in a very short time. This illustrates why the benefits and the time frame as well as the costs need to be quantified, e.g. the cleaning industry case study in Chapter 5A involved a significant investment in equipment but the pay-back period was short.

2.5.3 Assumptions about people as consumers

Consumers are the shoppers in society. They are considered to want to get the most from their money, today. They are assumed to know what items cost and be prepared to shop around to get the best price on items they want to buy, forcing prices to be equal everywhere for any particular product. Consumers are also assumed to have all the information they need in order to make an informed choice about their purchases.

But are these assumptions true? Stop to think for a minute about your own shopping habits.

- Do you travel from supermarket to supermarket buying a packet of cereal in one, milk in another and sugar in a third, or do you go to the most convenient supermarket for everything?
- Do you make an informed choice about buying a new car or just buy the same model and brand?

In reality the amount of time shoppers spend researching individual product and service prices and product quality relates to the value of the purchase. That is, more time will be spent researching major purchases.

It is also assumed that consumers are not concerned about how their purchases are produced and have no interest in internal enterprise practices, e.g. goods produced by child labour or from non-renewable resources. Consequently consumers would not be prepared to pay more for products in order to change these practices. These assumptions about consumers discourage enterprises from implementing any changes that would result in an increase in the price of their product(s). Do you think about your purchases and how they are produced and, if you do, what makes you think about these factors?

We believe that consumers are becoming aware about how products are produced. Far from following the economic assumption that consumers will not care how products are produced as long as they are as cheap as possible, consumers are demanding that products be produced in ways that benefit workers and the environment.

It is becoming common for foods to be labelled with specific information about how they have been produced. For example:

- following extensive publicity from conservation groups, methods of catching tuna have been changed to avoid catching dolphin as well. This is reflected on the labels of tins of tuna which carry a 'dolphin safe' symbol
- following lobbying from consumer groups, foods are increasingly being labelled to indicate if they contain genetically modified material.

In the clothing industry, a long-time bastion of sweated labour, publicity campaigns have aimed to encourage consumers to buy products produced under fair wage conditions.

Consumer buying habits are not immutable

Garment workers everywhere in the world today are faced with decreasing wages, deteriorating health and an increased risk of losing their jobs. The Clean Clothes Campaign aims to improve working conditions in the garment and sportswear industry. The campaign started in The Netherlands in 1990 at which time stores there were not taking any responsibility for the working conditions under which the clothes they sold were made. Things have changed and now there are Clean Clothes Campaigns in ten Western European countries. It is now more difficult to find retailers who denounce this responsibility. Campaigners are regularly in touch with organisations in a variety of countries, including those where garments are produced, and in this way work together as a network to draw attention to labour rights issues in the garment industry.

The Clean Clothes Campaigns in each country are coalitions of consumer organisations, trade unions, human rights and women's rights organisations, researchers, solidarity groups and activists. Every national campaign operates autonomously although they do work together towards international action, (Clean Clothes Campaign, 2002).

Campaigns show that, on the basis of information about production processes, consumers are prepared to break the economic assumption that price is the overriding factor for sales. However, it does appear that publicity ('information' in economic assumption terms) is the key to changing consumer buying behaviour.

2.5.4 Assumptions about people as workers

The contribution that workers make to the production process is termed *labour*. Economic theory makes some fundamental assumptions regarding labour, namely:

- *the labour of different workers is of uniform quality* so that no one person is preferred over another for any particular job
- *workers fully understand all the risks they face in their job*s and only accept employment if the wage rate is considered to compensate them for the risk of injury, i.e. workers are as safe as they want to be
- *the wages paid for their labour reflect workers' willingness to face those risks.* Workers know what employers are paying for labour and are prepared to shop around to get the best wage; this forces wage rates to be equal everywhere for any particular job
- *there is nothing to stop workers moving or travelling to anywhere to get suitable work*
- *workers will simply go to other jobs if the wages and conditions on offer are not good enough*; that is, they are not constrained to stay in unsafe or unrewarding jobs.

Knowledge and risks of asbestos miners

If the assumption about workers fully understanding all the risks they face in their job was true then it would be true to say that asbestos miners knew they risked contracting the disease mesothelioma and they knew and understood that mesothelioma could kill them. Furthermore, there was no need for governments to regulate asbestos mining because the miners were happy that the wages they were paid were sufficient to compensate them for the risk of chronic illness and death because, if they had been dissatisfied, they would have gone elsewhere or refused to work for those wages.

We now know that workers were not informed about the risks of mesothelioma and, because of the long time lag between exposure to asbestos and the occurrence of mesothelioma, they did not understand the risks they faced. In addition, these workers had families and social support in the local towns making it difficult for them to move elsewhere. Their skills were based around asbestos mining so employment opportunities elsewhere were limited for them.

On the basis of the dangers from asbestos and uneven knowledge and power between workers and asbestos mining companies a case was made for government regulation to protect workers' health and safety.

These may be reasonable assumptions for some types of workers, e.g. young skilled workers whose skills are in demand, but they will not always seem to be reasonable.

You can probably think of many less obvious situations where some or all of these theoretical assumptions about employment would be untrue. For instance:

- families cannot move freely as then they leave behind their social network, moving disrupts children's schooling and moving itself costs money. In the north of England and Wales, due to closure of the coalmines, miners have been unemployed. Successive governments have urged them to move to the south of England where there is employment, ignoring the fact that the jobs in the south require a completely different set of skills, housing and food are more expensive and their families and friends are 'back home'
- workers with low job skills are less able to be choosy about accepting jobs. A manual labourer may know he is at risk of back injury but not have the literacy skills, education or experience to be able to move to less physically demanding work
- sometimes workers will not know, or may not fully understand, the risks involved in their jobs. For instance, health care workers exposed to chemotherapy drugs may develop illness in the long term but they do not necessarily follow safe procedures to prevent exposure (Gibbs, 2002). Young workers often express the opinion that 'it won't happen to me'
- workers will often decide that having bread on the table today is more important than any possible risk of injury or death. For instance, workers may be constrained to continue working in unsafe conditions by lack of employment

alternatives, lack of social services or having no savings to support themselves while looking for alternative employment

• in times of high unemployment enterprises can offer workers a 'take it or leave it' ultimatum, which will lead to more workers accepting higher risks.

Workers are seen, then, as a cost to the firm and most accounting models do not have the capacity to record the skills and experience of workers as an asset to offset these costs. Some enterprises that produce knowledge and information are starting to change, but skills and experience of workers are usually not considered in economic analyses.

2.5.5 *Assumptions about production decisions*

The producers of products and services (this group also includes the owners and share-holders who invest in the production process) are considered to have as their fundamental goal the maximisation of current profits. All decisions within the enterprise, whether they be about the hire of workers, the purchase of equipment or the pricing structure of products and services, are assumed to be based on this one goal.

It is further assumed that enterprises can adjust production of services and products by increasing and decreasing the labour force without penalty or delay. Employment contracts make the adjustment of work hours less flexible than the simple assumption, although the assumption is partly supported by the use of casual workers (precarious workers), fluctuating shifts and overtime. By talking in general about the labour force and keeping the discussion in general terms, the ethics of the impact on workers and their families is kept out of the decision-making process.

These assumptions validate the adjustment of employment levels through such processes as 'down-sizing' (particularly sacking full-time workers) and the increased employment of part-time and casual workers. Where a worker is permanently disabled through workplace injury or disease they are replaced at the workplace although many of the costs of disability are borne by the community: the enterprise is subsidised to the extent that it does not fund these costs. Economic models do not consider there to be any kind of social contract between workers and enterprises (see section 2.4.1 – production and supply of products and services).

Economic theory assumes that an enterprise will need to make choices about how it spends its money and where it diverts its workforce. The theory assumes that not all the projects or activities that the enterprise would like to do can be funded at one time because money is in limited supply. Let's face it, if there was no restriction on money or workers then there would be no need for extensive funding proposals or budgets to justify expenditure.

Decisions made by producers are based on the present information about future sales. There are no 'crystal balls' to provide exact information about future sales, and producers can rely only on the information available to them at the time they make their decisions. To a greater or lesser extent the roles of marketing, finance and accounting all involve a degree of 'crystal-ball gazing' to estimate these future sales and prices.

Adjustments will be made to production schedules based on readily available data such as stock in hand and forward sales. For example, as stock in hand rises and forward sales diminish, production will be slowed and vice versa, which is a 'negative feedback' situation.

As Keynes (1936) put it, 'it is sensible for producers to base their expectations on the assumption that the most recently realised results will continue, except in so far as there are definite reasons for expecting a change'. Thus, in a very real sense, producers (as economists) walk backwards into the future.

Having determined its production goals, the enterprise needs to consider how best they can be achieved and this is where economic analysis may be brought in to help the decision-making process. Economic analysis is assumed to provide an unbiased answer to the question posed but, as we have seen, the answer also relies on the underlying economic assumptions.

2.6 How to use economic analysis as a decision-making tool

Economic analysis methods can be used as a *tool* to evaluate the financial viability of funding proposals; that is, the costs versus benefits of alternative proposals.

If you are to use a model (for example, the Productivity Assessment Tool) you need to know, and perhaps even defend, the assumptions on which you predicate your economic ideas. You may wish to refer back to section 2.5 when you are writing up your proposal to remind yourself of the assumptions underlying the analysis.

The principal concept of economic analysis is that if a project returns greater benefit to the enterprise than it costs, then it is worthwhile. Its approach is to present cost data and benefit data numerically based on estimates of future events, and weigh the costs against the benefits. Economic analysis can never provide an exact answer to any question because no-one can look into the future and know exactly what will happen. It is merely one possible tool to guide the decision-making process, but judicious use of these tools can turn decisions in your favour.

In our day-to-day work we are often called upon to make judgements that are based on our best estimate of what will happen. For instance, making a judgement to employ a particular person is based on an estimate of their future performance in the job and in the enterprise. Engineers making a judgement to purchase particular equipment or systems base their advice on an estimate of how it will perform in their enterprise. It stands to reason that an experienced human resources manager, occupational health and safety manager or ergonomist will be as good at developing estimates in their area of expertise as an experienced engineer is in their field.

For inclusion in economic analysis your estimates need to be presented quantitatively. For example, experience may say that 80% of new workers pass their initial trial period successfully and continue in employment with your enterprise. When estimating a budget for recruitment costs, the human resources manager would be likely to request funds to cover 120% of the estimated vacancies for any financial period to account for this drop-out rate. The estimate would be developed in the same way if the human resources manager were including recruitment costs in an economic analysis instead of a budget.

The numbers that are used in the economic analysis should reflect the best estimates by experienced personnel, who may include managers, workers, and/or other enterprises operating the same systems/machinery/technology. Barring unpredictable outside events, such as fire or storm damage or a stock market crash, these estimates will most likely be reasonably accurate.

Having decided what factors to include in the model and having calculated or estimated the value of each item, the answer is simply arrived at by mathematical calculation. The benefits are added together and the costs subtracted from them; if the answer is positive then the benefits outweigh its costs and the project is worthwhile.

Other data included in the economic analysis will reflect economic assumptions; that enterprises want to be profitable and make the most profit they can now. On this basis if you are comparing alternative proposals, the proposal that has the greatest benefit compared with costs, including a time factor, is considered the most worthwhile. This is a sensitivity analysis.

You need to act cautiously when considering your 'answer' and acknowledge the assumptions you have made. The figure you have developed from your calculations is only going to have meaning if it truly reflects the real situation, as best you can achieve; the old saying is '*the devil is in the detail*'.

When you look at someone else's proposal you can never assume they will have taken into account all the factors that you consider relevant. Safety costs and worker productivity costs will be left out of the calculations if they are:

* not seen to be relevant to the decision-making process
* perceived to be too difficult to quantify
* not realised or understood to be important.

In Chapter 4 we show how the Productivity Assessment Tool is designed to ask the relevant questions to elicit appropriate cost data. Even so, you will find situations where ethical and personal judgement are required to arrive at your data. Consider the following:

> *Question*: if there has never been an injury from an *existing* machine, or system of work, but it is considered to be dangerous to the point where a severe injury is likely, do you include an allowance for injury costs and, if so, what figure should you use?
>
> *Answer*: your workers' compensation insurer or industry association should be able to provide you with an average value (or a range of values) for the type of injuries you think may occur. Whether or not the inclusion of this figure in an economic model is acceptable to management is another issue and needs to be answered on a case-by-case basis.
>
> *Question*: installation of a *new* work system raises potential health and safety effects. How do you establish and cost these unknown effects?
>
> *Answer*: contact the manufacturer or supplier or try to locate an enterprise that has installed the system and ask for their experience. Otherwise you could undertake a risk assessment and base your estimates on the results. All cost and benefit figures will be based on estimates.

Within your enterprise you will need to make judgements about the potential costs and benefits for each item you consider important in your economic model. The Productivity Assessment Tool will assist you to select the important items, based on worker costs, for short-term projects. However, if your analysis needs to consider a long time period you will have to include estimates for costs that are likely to change over the duration of the project, for example, potential wage increases.

Economic models are also increasingly being used to evaluate the worth of alternative proposals for large-scale enterprises or government spending. The documents produced often run to hundreds of pages using complex mathematical calculations but, even here, it is important to apply the fundamental principle of reasonableness of the assumptions. You can look at the assumptions that have been used to build the model and ask:

- *do the assumptions make sense?* Check whether the values included for each estimate seem reasonable and whether the model has included all (or at least the major) factors that should be considered. If there are major factors missing or unreasonable estimates then the conclusions will have less credibility than otherwise
- *if you recalculate using other assumptions do the results change?* The assumptions may bias the answer in one direction or the other; so does changing the assumptions have a significant impact on the result?
- *does the type of model chosen suit the question to be answered?* It is important to choose the most appropriate type of economic model and the following sections discuss this question.

2.6.1 Types of economic analysis models

There are three main types of economic models that can be used to analyse the costs and benefits of a proposal. These are, from the simple to the more complex:

- financial appraisal – return on investment analysis
- cost-effectiveness analysis
- cost-benefit analysis.

A description of each of these models follows together with their relative strengths, weaknesses, and an example of how they are used.

Financial appraisal

Financial appraisal or return on investment (ROI) is the simplest of these models. It only considers costs and benefits that affect the enterprise itself and is a microeconomic model.

Factors that lie outside the enterprise are not considered, even where they may be considerable. For instance, pollution of a local river would not be considered as a cost to the enterprise unless the enterprise were legally or contractually obliged to clean it up. In the same way the illness cost from long-term occupational disease of workers would not be a cost unless the enterprise were legally or contractually obliged to meet the workers' health costs.

This model considers the cost of a proposal first. These costs are often termed an outlay or investment. It then calculates the benefits from the proposal and takes into account the time frame which it will take to recover the initial outlay or investment, hence the term *return on investment*.

Cost-effectiveness analysis

Cost-effectiveness analysis compares the costs and benefits of a proposal to the enterprise, including the social and cultural, and is a macroeconomic model. This type of analysis is useful where factors external to individual enterprises need to be included (e.g. the environmental impact of mining, the impact on the transport infrastructure of building a new suburb or the long-term health effects of chemical exposure) or where future costs or benefits cannot be reasonably quantified.

Cost-effectiveness analysis is commonly used in analysis of alternative proposals within the government sphere. For example, in public health or occupational health and safety projects, where the value of lives saved needs to be factored in, cost-effectiveness analysis poses an alternative to the need to place a dollar value on human life. Using cost-effectiveness analysis, a health-screening program for early detection of cancer can express its program costs in terms of the cost per estimated number of lives saved, rather than placing a monetary figure on the value of the lives saved.

The limitation of the model is that the answer provided by the analysis still needs further analysis and decision-making. Its strength is that it does not force values to be placed on poorly understood factors in the calculation. Cost-effectiveness analysis can be used where the dollar value of the benefits cannot be reasonably estimated. Instead of costing the benefits of the proposal they are expressed in terms of the cost per unit to achieve them.

Cost-benefit analysis

Cost-benefit analysis is the name commonly applied to *any* economic analysis which examines both the costs and benefits of a proposal. In order to be unambiguous we have chosen to use the term cost-benefit for this specific economic model. The Productivity Assessment Tool is an example of a cost-benefit model.

Cost-benefit analysis refers to a model that purports to identify all the costs and benefits associated with the proposal under consideration, including society-wide costs and benefits. This is an assumption in itself, as it is not feasible for all costs and benefits to be known. Estimates of the known costs and benefits are quantified in monetary terms and then directly compared with each other. A proposal is considered to be worthwhile when the benefits of the proposal outweigh its costs. Where two or more proposals are directly compared, the project with the greatest *net benefit* is considered the most worthwhile.

Cost-benefit analysis is the most complete model as it attempts to include the social and cultural aspects of any proposal. This type of analysis is useful where factors external to individual enterprises need to be included, e.g. environmental impact of mining; transport infrastructure effect of building a new suburb; the effect of new occupational health and safety regulations. For this reason it is the type of modelling most commonly used in analysis of alternative proposals within the government sphere.

The limitation of the cost-benefit analysis model is also its greatest strength. Placing a dollar value on all factors results in a clearer, simpler model that is unambiguous and easier to interpret. Unlike cost-effectiveness analysis, once the final figure is calculated there is no need for further interpretation and, unlike financial appraisal, an allowance can be included for the social costs and benefits of the proposal.

The differences between these three models (financial appraisal, cost-effectiveness analysis and cost-benefit analysis) are illustrated in example 2.2.

Example 2.2: Comparison between the three economic models

Fresh Supermarkets wanted to assess the viability of replacing their current flat bed trolleys with electrically powered height-adjustable trolleys. They have three supermarkets in Lilyfield which are each open for 360 days per year. Each trolley costs $5,000 so their initial outlay would be $15,000. They estimate that the new trolleys would save them 20 minutes in shelf packing time per store per day. They pay their workers $15 an hour and do not include overheads and other costs when developing investment proposals.

Calculation of financial appraisal

The return on their investment (ROI) is calculated as the number of days it would take to repay the outlay for the three trolleys.

Given that the number of supermarket opening days per year is known (360), the return on investment can be recalculated in terms of the time it would take to repay the investment. For the saving of $15 per day in wages and a cost for the trolleys of $15,000 the repayment time is 15,000/15 = 1,000 days or 2.8 years (1000/360) (see table).

Savings in time and money for the three stores	60 minutes per day or $15 per day in wages for the three stores
Cost of three trolleys	$15,000
Return on investment (ROI)	15,000/15 = 1,000 days = 2.8 years

In most enterprises management sets targets for the time frame in which investments should be recouped by the enterprise. In stable and secure enterprises this time frame is usually longer than in new and/or unstable companies; whether or not 2.8 years would be considered a reasonable investment in this example would depend on management's concept of a fair return.

Cost-effectiveness analysis

Fresh Supermarkets have never had any injuries among their packers and are not sure whether the purchase of the new trolleys will prevent some future injuries; they are uncertain whether an amount should be included to represent potential injury cost savings, and if so, how much. Fresh Supermarkets have chosen to use cost-effectiveness analysis in this situation as the savings to be gained from introducing the new trolleys do not need to be quantified in terms of potential injury savings.

There are eight packers in each store who will use the new trolleys for stacking shelves. This is a total of 24 workers in the three stores.

To calculate the cost-effectiveness of the proposal, the total outlay on the trolleys is divided by the total number of workers who stand to benefit (see table).

Cost of three trolleys	$15,000
Number of full-time equivalent workers who will benefit	24
Cost-effectiveness of investment calculated as the cost for each worker who will benefit	15,000/24 = $625 per worker

In this example the cost per worker is a proxy for the value of the total benefits, including benefits that can be quantified and those that cannot.

On the basis of economic theory, proposals with a lower cost per worker should be funded first. The management team will also receive requests to fund other proposals and whether or not this project is funded depends on these other proposals – are they at a lower cost per worker for the same perceived benefit?

Cost-benefit analysis

Although Fresh Supermarkets have never had any injuries among their packers, they discussed the situation with their industry association and decided that it is likely there will be an injury if they do not change to the new trolleys. Between them they developed an estimate of the cost impact of an injury at work. They decided to use the mean direct cost of a back injury in their industry, which is $11,000.

To calculate the cost-benefit (the pay-back period) of the proposal, the total cost of purchasing the trolleys is divided by the estimated benefits for a one-year period. The benefits include the estimated saving for a back injury ($11,000) and the increase in productivity, which is the time saved during the one-year period (one hour or $15 per day). As the supermarket is open for 360 days per year, the increase in productivity would be $15 \times 360 = \$5,400$ (see table).

Total cost of trolleys	$15,000
Total benefit of trolleys/year (back injury savings plus productivity increase)	$11,000 + $5,400 = $16,400
Cost-benefit (calculated as the pay-back period)	$15,000/16,400 = 0.9 years (11 months)

In this case the pay-back period is only 11 months.

On the basis of economic theory, proposals where the benefits outweigh the costs should be funded, with those providing a greater benefit being funded first. In these simplified examples, although the return on investment calculation gave 2.8 yearsand the cost-benefit model gave 11 months, neither is right nor wrong, correct nor incorrect; they depend on the assumptions made and the factors considered. Where only the cost of the trolleys is considered, without factoring in the potential benefit to workers, the purchase of trolleys takes much longer to pay back the initial outlay.

2.6.2 Differences between short-term and long-term analysis

The Productivity Assessment Tool is a cost-benefit analysis model that assumes the money is spent and the benefits are gained within a short period; in our experience most ergonomics interventions have a pay-back period of considerably less than one year.

Where it is calculated that costs or benefits will not be contained within a short period, the costs of the project are increased to account for interest rates (the cost of money) and benefits deemed to flow from the project are discounted by an amount that accounts for the need to wait to enjoy them.

The following sections discuss short-term and long-term projects.

Short-term models where the value of money is unchanged

In short-term models it is assumed that the value of the money used to fund the proposal will remain approximately the same over the course of the project and it is usual to use the input data and resultant data without modification.

The Productivity Assessment Tool has been developed to consider the costs and benefits of a proposal where they occur within a relatively short term. The basic version of the Productivity Assessment Tool is included in the CD accompanying this book and is discussed in Chapter 4, with examples of its use in Chapters 5 and 6.

Long-term models incorporating changing value of money

Where the costs and/or the benefits of a proposal are calculated to take several years, or if you are working in an environment of hyperinflation where the value of money changes at a rapid pace, you may need to adjust the net costs/benefits of the project. Costs may be increased to account for inflation and interest rates, and benefits may be discounted to take into account the need to wait to enjoy them. In either case additional calculations would be needed. In the CD accompanying this book you will find a program to assist you to calculate changing values.

The rationale for increasing costs to account for interest rates or inflation is that if the money were available now you could spend it on another money-making venture and you would then have more money. This harks back to the fundamental assumption about the behaviour of enterprises that they want to maximise profit today. Also enterprises may need to borrow the money to fund the project and then have to pay back both the capital and interest.

The rationale for reducing benefits where the benefits are not seen for some years is that the worth of the benefits that will accrue in the future will be eroded by inflation. There is no rule about how much discount should be applied to future events. As the concept is to reflect the amount of money the enterprise could otherwise have earned, and to account for inflation, it is usual to use a figure such as the interest rate available on the long-term money market. The term 'discount rate' is used for the percentage discount to be applied.

Some governments and enterprises set a rate that they use for the discount rate as a matter of policy and this may be higher or lower than the long-term money market rate.

The rate set can have a profound effect on the viability of a proposal for long-term expenditure or investment. The higher the discount rate chosen for the project, the less likely the project is to return a positive benefit as the net benefits (benefits – costs) are discounted exponentially over time. This will be of particular concern if you are seeking funding for a project that will take several years before it starts to show any net benefits; such projects include prevention of occupational cancer from hazardous substances or radiation exposure, prevention of industrial deafness, implementation of large scale multi-skilling, or remediation of land contaminated by chemicals. Example 2.3 illustrates the use of discount rates for a long-term proposal.

To calculate the cost-benefit analysis for a long-term project it is usual to estimate the costs and benefits on a year-by-year basis, then subtract costs from benefits to obtain the *net* benefit. Benefits are calculated for each year for which there will be benefits gained and some projects, such as cancer prevention or environmental projects, can accrue benefits over decades or even centuries. Costs are calculated for each year that the enterprise will invest money to implement the project: this is often a shorter period than the time over which benefits will accrue from the investment. The values for benefits can be discounted and/or values for costs can be increased to account for the change in the value of money over the life of the project. The net benefits accruing from the project each year are then added together and this final figure is given the term the *net present value (NPV)* of the project.

A regulatory impact assessment, or any other macroeconomic analysis, should clearly state the discounting value/interest rate it has used to calculate the net present value of the project and the reasons for the chosen value(s). The CD included in the book enables calculation of NPV.

The NPV can be calculated using a formula, which is written as follows:

$$\text{NPV} = \sum_{n-1}^{0} B/(1 + r)^n$$

The formula may also be written:

$$\text{NPV} = B + B/(1 + r)^1 + B/(1 + r)^2 + B/(1 + r)^3 + B/(1 + r)^4 \ldots + B/(1 + r)^{n-1}$$

where B = annualised value of the net benefits (benefits – costs), r = the discount rate, and n = the number of years for which benefits will accrue from the project.

It is worth emphasising that if the savings to be made from your proposal will not occur for some years then their value will be significantly reduced when discounting is applied.

If you decide to use a discount rate, ask yourself:

- *is the discount rate reasonable?* You can check the long-term bond rate to give you a guide. This reflects the stock market's best guess about how the economy will perform over the next 10 years, including expected interest rates and inflation
- *does the discount rate significantly impact on the project's expected benefits?* For instance, is there a lag between the intervention and the outcome?
- *are there other issues that should be considered?* Just because you can model the impact of an intervention does not make the model right or just. Consider the ethical issues involved in the decision-making, as economic models are only tools to help give a financial structure to your argument.

Example 2.3: Impact of discount rates on pay-back period

If you were looking to spend $100,000 on a project that you expected to return $10,000 per annum in net benefits over 10 years without discounting, you may expect the proposal to have a 10-year pay-back period (see Figure 2.1).

However, when a discount is applied to these net benefits, the net benefits returned over the ten-year period are reduced or *discounted*. Discounting is calculated like compound interest in reverse. Instead of the benefits growing as time passes, they reduce.

Figure 2.2 shows the impact of discounting the $10,000 per year benefits by 5% and 10% over a 10-year period. As successive years pass the net benefits returned from the project continue to fall away. From the table below you can see that it would take 15 years to pay back the initial investment at the 5% discount rate, and at the 10% discount rate the initial investment would never be repaid.

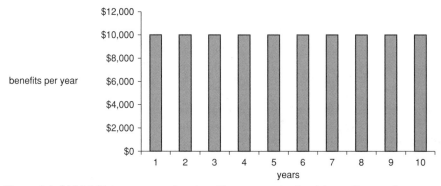

Figure 2.1 $100,000 investment showing 10-year pay-back without discounting.

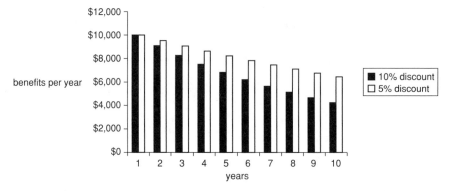

Figure 2.2 $100,000 investment showing impact of discounting on future benefits over 10 years with 5% and 10% discount.

Discount applied	Total-value of the net benefits over 10 years	Pay-back period for the project
0%	$100,000	10 years
5%	$81,078	15 years
10%	$67,590	>100 years

When you change the inputs into a cost-benefit model you will change the answer. Some changes will have a greater impact than others and, if one is to make an ergonomics intervention, it is just as well to identify those parameters that will lead to the greatest cost-benefit. That is, if you have to 'sell' the intervention by means of it being cost-effective then you will need to know that the intervention will tackle the most significant cost parameters.

For example, if you find that overtime is a major cost factor it may be better if your intervention reduces overtime than, say, marginally increasing productivity on the assumption that in ergonomic or injury prevention terms they are equivalent. A sensitivity analysis will assist you in this and is dealt with in more detail in Chapter 4.

2.7 Developing your proposal

Having undertaken an economic analysis and satisfied yourself that the project is worthwhile, it may be necessary to undertake a further step before applying for approval and resources. The way you present your proposal should:

* match the complexity of the funding decision process
* enable direct comparison with other proposals that are being compared against yours.

If you are competing for funds against other sections that have put forward detailed proposals justifying their request for money, your proposal will require a similar depth of information or detail. If you are explaining your proposal in an informal meeting you may simply need to use some of the terminology of economic analysis and show how the project is worthwhile to the company's bottom line.

Other proposals being considered are likely to have been prepared using an accounting model. As discussed in section 2.5, accounting models do not cost workers in a way that occupational health and safety/human resources systems do, and can ignore the value of workers. To enable a direct comparison between proposals prepared using the Productivity Assessment Tool and those based on an accounting tool it may be necessary to point out these differences, perhaps recalculating the costs of alternative proposals giving consideration to the value of workers. Alternatively it may be necessary to include justification for including the value of workers in your proposal (see section 2.5.4).

In the following chapters we will show you how you can present your data using the Productivity Assessment Tool. Because the model automates the calculations it makes it quicker and easier to prepare data for presentation in your proposal.

2.8 References

Adams, D., 1979. *The Hitch Hiker's Guide to the Galaxy* (London: Pan).

Clean Clothes Campaign, *News Releases March 7 2002*, http://www.cleanclothes.org.

Department of Foreign Affairs and Trade, 2000. *Review Of Australia's General Tariff Arrangements: Submission to the Productivity Commission* (Canberra: AGPS).

Farrell, C., 2001. *Dismal Science, Woeful Record, Useful Insights*. Businessweek Online, 29 June.

Gibbs, S., 2002. Safety is a fuzzy word, *Ergonomics Australia* 16, no. 1, 8–39.

Keynes, J.M., 1936. *The General Theory of Employment, Interest and Money* (London: Macmillan).

McClintick, D., 2001. The Decline and Fall of Lloyd's of London. *Time Europe* **155**, 7.

Manning, I., 1998. *Economic Models: What Can They Really Tell Us?* (Sydney: Public Centre Research Centre, University of NSW).

NEIR, 2000. *Industry Assistance Reductions and the Australian Economy: Contemporary Issues and Future Prospects* (Canberra: AGPS).

Samuelson, P., 1980. *Economics*, 11th edn (New York: McGraw-Hill), p. 2.

Selikoff, I.J. and Lee, D.H.K., 1978. From *Encyclopaedia of Occupational Health & Safety*, 4th edn (Geneva: International Labour Office).

3 Information sources

Contents

3.1 What this chapter is about

When stepping into a workplace to make a cost-benefit or workplace assessment we are frequently met with the question 'where do I find the information?', but this is the wrong starting point.

The very first question should be '*what* questions should be asked?' The pertinent questions (those that indicate the important parameters of the work area) will lead to the information (the data) you need. Questions must come before answers, which is logical, and thus the framing of the questions is the first and most important task. The value of the Productivity Assessment Tool, included with this book, is not simply its calculating power but, far more importantly, pointing you in the right direction – posing the right questions. The questions that should be asked will be related to:

- the people involved
- their work methods and practices
- their equipment.

This chapter helps you to define and ask the pertinent questions.

3.2 Defining the work area

It is essential to the working of the Productivity Assessment Tool, or for that matter any other cost-benefit analysis model, to define carefully the work area which will be subject to analysis.

The work area will be defined by:

- similarity of the tasks performed
- those people who will be affected by the intervention.

It cannot be stressed enough that the work area chosen for the cost-benefit analysis must be completely and rigidly defined. Ask the question 'what will change through the intervention?' as that will help define the work area.

Cost-benefit analysis is a time difference concept, and if there is no difference in the *before* and *after* situations for any particular cost item then the end parameters (pay-back period and savings) will not change. If you are looking at reducing back injuries in the warehouse staff, to include the office staff in the work area will add nothing to the effectiveness of the intervention for the warehouse staff. It will complicate the calculations and act as a distraction by deflating some of the benefits.

A common error, and one to be avoided, is to change the defined work area in the before and after situations; e.g. defining the work area as the warehouse staff (before situation) but including the costs for improvements both for the warehouse staff and the office staff. One should only compare apples with apples, not with oranges.

If there are several production lines and only one is to be changed, then the work area is that production line only. If the people change from line to line, then the number of people on the intervention line must be assessed by the total hours worked on that line; if, on the average, four people spend two hours each per shift on the intervention line the line requires one full-time person (for an 8-hour day). On the other hand, you can say that it takes four part-time people (each working two hours per day) to run the intervention line. Either way is valid and is accommodated within the Productivity Assessment Tool.

It may be necessary to include data from outside the strict confines of the chosen work area. For example, if supervision is likely to be altered by the intervention you need to include supervisory costs and if, for example, labour turnover is likely to be altered then you need to include an allocated or proportionate cost for the personnel or human resources department.

The relevant questions to help you define the work area are:

- what information do the decision-makers require?
- what will change through the intervention?
- am I collecting unnecessary data that will not change through the intervention?
- how many people work there (based on total working time)?
- what costs or activities do we include from outside the workplace that will change?

Although it is important to define your work area rigidly you may need to bend your definition to include information the decision-makers want. For example, if they will not be satisfied unless you include the office staff in your calculations of reducing back injuries in the warehouse staff, then you will have to include the office staff. In reporting your results you will need to explain how those workers who are not affected by the intervention will dilute the apparent effectiveness of the intervention.

3.3 What questions should be asked?

After defining the work area we must ask the pertinent questions. These questions are (1) the basic questions required for any cost-benefit calculation and (2) the specific questions concerning the intervention.

(1) The basic questions are:

- numbers of workers
- hours worked
- direct wage costs.

(2) The specific questions concerning the intervention include:

- overtime
- equipment
- productivity
- supervision
- administration
- recruitment
- intervention costs.

Once you have started to address each specific question you will soon see which are relevant to the intervention and which are not. For example, if there is no labour turnover in the defined work area then collecting data on recruitment costs will have no relevance to the value of the intervention. On the other hand, if there is a high labour turnover and the intervention is expected to reduce turnover, then the questions that follow will point you in the right direction (questions of training, productivity, extra administrative and recruitment costs, for example).

3.4 What accuracy is needed?

Even before gathering data one must be aware of the *accuracy* required of the data. It is said that 20% of the time is spent collecting 80% of the data; after this one must assess whether or not it is worth spending the extra time, 80%, to collect the remaining 20% of data; it rarely is. It would not be worth spending extra time when the pay-back period is short, but for long periods then a higher order of accuracy and a more complete collection of data may be required. In our experience most ergonomics interventions lead to a short pay-back period and thus high accuracy is often not required.

For example, if the calculated pay-back period for a particular intervention is four weeks, even if your data is inaccurate by 50% the actual pay-back period will be between two weeks and eight weeks. In any event, the pay-back period is so rapid that the inaccuracy does little to change the conclusion that the intervention is well worthwhile. On the other hand, if the pay-back period is calculated to be three years and your data is only accurate to 50%, the actual pay-back would be between 18 months and six years and one would look to a higher level of accuracy. As a guide it may be advisable to collect the readily available data and do a rough calculation.

Once you have done your first, rough calculations you will be able to see if you have asked the pertinent questions, whether the data you have is relevant and to the level of accuracy you need.

So, for the cost-benefit analysis you will need to:

- define the work area where the intervention will be effective
- determine the pertinent questions
- seek the data that are relevant
- be aware of the accuracy or limitations on the data.

It is also useful to know that:

- you can never have all the data (logically, one cannot know whether there are any more data)
- different people (workers, managers, accountants, etc.) will have a different perspective on what is, or is not, important
- data can be impressive but actually worthless, as it may not show the essential elements.

In the final analysis the questions you ask will determine the data you collect and influence the way you wish to present the cost-benefit analysis.

With these mild warnings, what are the relevant questions? The rest of the chapter is concerned with both the pertinent questions and the way to collect data.

3.5 Easy-to-collect data

Data concerning

- wage costs
- paid absences
- overtime
- supervision
- recruitment and administration

can be found quite simply from the accounting and payroll systems. These sources will give routine information about direct costs and the indirect or overhead costs: wages; sick leave; injury absence; overtime; labour turnover; insurance; tax and other obligatory costs; supervisory and administration costs. Mostly the data will be reasonably accurate, and certainly accurate enough for most analyses.

Indirect and overhead costs data can be included to increase the accuracy of the determination of the total cost of employment and will be important in determining the resources that it is 'worth' (in the economic sense) employing. For example, if only the directly paid wages are used as the cost of employment this will be an artificially low amount as the actual employment costs include insurance, vacation and other paid absences, supervision and so on. These other costs will double or even treble the wage costs and give a better understanding of the cost of employment.

With the full cost of employment known, the benefit of any intervention will better reflect the employment situation and the financial value of the intervention.

Some of the data may only be available for the entire enterprise so you may need to take a proportionate or allocated cost. For example, if there are 100 employees and ten of these work in the defined work area where the intervention is to take place, then you can add one-tenth of the administration costs to the defined work area. On the other hand, some enterprises just say that overheads are $x\%$ of wages; although the accuracy in these cases is not usually as high, using this figure will make your analysis comparable with other proposals.

The indirect and overhead data will present a more complete picture of the costs of employment and be a more realistic, and perhaps more acceptable, way to look at these costs; however, if the indirect and overhead costs do not change in the *before* and *after* situation these costs will not affect the pay-back period.

If this method of looking at the cost of employment is used across the entire enterprise, other proposals will be prepared on the same basis: if you have inaccuracies in your data, your colleagues will have the same inaccuracies in the data for their proposals.

3.6 Performance data

Performance data includes training time; error rates; warranty and repair costs; waste and energy costs; down-time of equipment; and output, both qualitative and quantitative. If data is only available for the entire workplace, the data required for the intervention may need to be estimated or re-calculated. For example, if the intervention will have an effect on energy use in the defined work area but figures on energy use are only known for the entire workplace, they will need to be estimated for the machinery in the intervention (defined) work area.

What is included and what is not may also be a matter of enterprise conventions. Some companies do not recognise that product warranty costs should be a cost against the work area where the errors occurred or that a proportionate cost for down-time should be allocated to various work areas. If these cost parameters change with the intervention then they are critical values and need to be included; excluding them will disguise or dilute the value of the intervention. If this is the case then you must argue for including these costs, perhaps by understanding why they were excluded in the first place.

3.7 Quality data

Quality is an important factor of productivity. Productivity does not only mean *quantity* of output; *quality* is usually as important. It is no secret that many Western manufactured cars were of poor quality until Japanese manufactured cars, of higher quality, came onto the Western market. It did not take Western manufacturers long to realise that number of cars made per day (quantity) was of little importance if nobody wanted them, preferring Japanese cars of a superior quality.

Thus measures of quality must be added to those of quantity when assessing an intervention that will affect productivity. Deming (1982) has pointed out that defects in manufacturing production are not free and that the persons making the defects get paid as well as the persons who are correcting the defects. Product warranty costs are a common measure of quality; if a customer brings back the item for replacement or repair that sale may become a net loss rather than a profit.

The cost of poor quality

One of Deming's (1982) concerns was that manufacturers should make a product only once and make it free of defects:

> A plant was plagued by a huge amount of defective product. 'How many people have you on this line for rework of defects made in previous operations?' I asked the manager. He went to the blackboard and put down three people here, four there, and so on – in total, 21 per cent of the work force on the line.
>
> Defects are not free. Somebody makes them, and gets paid for making them. On the supposition that it costs as much to correct a defect as to make it in the first place, then 42 percent of his [the manager's] payroll and burden was being spent to make defective items and to repair them.
>
> Once the manager saw the magnitude of the problem and saw that he was paying out good money to make defects as well as to correct them, he found ways to improve processes . . .

Training and quality

Training will often have an effect on quality as the following example illustrates: The introduction of word processors to the publishing industry resulted in journalists directly setting the typeface rather than specialist typesetters (linotype operators). Needless to say, there were many industrial difficulties in this and the industry generally was plagued by dispute and by serious health problems, particularly upper limb injuries, among journalists.

In a particular magazine company the editorial areas were thoroughly overcrowded and the traditional method of ergonomics improvements to the work station, other than chairs, was not an option – they had to make do with the small old desks they had had for years.

This company decided that a good training regime might overcome some of the journalists' worries and that the training program had to treat the journalists as intelligent people – the reasons *why* one takes certain steps to overcome potential adverse health effects were as important as the means to do so.

The training consisted of groups of up to five people for one hour with a trainer well versed in anatomy, physiology and ergonomics so that he could illustrate and discuss the concepts from a professional point of view. He could be 'interviewed' (sometimes 'grilled') by the journalists so that they could get precisely the information they wanted. A desk and equipment (as well as a skeleton) were in the room so that each person could be individually trained.

One year later, not a single person had an upper limb injury (an industry first in this large city) and there was no decline in the quality or sales of the magazines.

In a call centre/software help desk, if the information provided to the operator is not up-to-date or is incorrect and this is passed on to the customer, then considerable errors may creep into the customer's operations – not a situation ensuring trust and continuing sales.

An intervention that includes training will often have an effect on the quality of the product or service and, if so, quality measures should be included in the cost assessment.

3.8 Spoken information

For many companies, especially smaller ones, there is often a lack of written information concerning productivity. The reason for this may be:

- lack of time or skills to collect relevant information
- lack of ability to analyse such information
- that relevant information is kept in the head and not committed to paper (or computer).

Whatever the reason, management, and especially employees involved in the intervention, will be an invaluable source of information. Information should not be discarded because it is not written down or does not carry as much accuracy or detail as one would like. In many instances, the workers will have very accurate information in their heads.

For some productivity data you may have to rely on inspired guesses; do not disregard inspired guesses. Clearly, one has to sift information carefully to ensure that exaggerations do not creep in but generally the information is good and, in any case, this may be the only source.

Agreed information

One of the authors remembers being asked, at a meeting of safety officers, when he is visiting a workplace as a consultant how he would determine the changes that should be made. After telling them that 'professional secrets would not be divulged', they were then told that the method is to ask the people in the workplace for their views, write them up in a report and present it to management along with the invoice! The point of this is that if all parties, management, supervisors and workers, are agreed on the data then it becomes the best reasonably available.

Much can be gained from the opinions of workers and supervisors regarding lost production due to faulty equipment, wrong tools, wastage, software errors and other productivity or service issues. Such information may be in the form of a proportion or a percentage (i.e. one-quarter loss; 10% increase) rather than in absolute figures but is, nonetheless, valuable information and easily converted into data useful for a cost-benefit analysis. For example, if the information is that wrong tools lead to a 10% loss of production this can be entered into the data screen for 'Reduced Productivity' in the Productivity Assessment Tool, which converts the figure to a cost of employment.

3.9 Questionnaires and checklists

Once you have gathered the relevant and easily available written and spoken data you may find that there are still essential gaps in your data. Checklists can be a useful data-gathering tool to fill these gaps.

Usually checklists are not specific to any one workplace but ask general, qualitative questions. Typical questions include:

- are loads lifted?
- are hands held above shoulder height during the task?
- has the person been trained to do the task?
- are the tools suitable?
- is the work station organised so that the person does not have to bend?

The answers to these general questions will indicate where, in the work cycle, the workers feel stress or discomfort and/or where there may be lowered productivity. This may enable you to focus the direction your intervention needs to take, but will not yield data that can be used in a cost-benefit analysis.

Questionnaires can be used to assess the employees' views both before and after an intervention (social questions) and are increasingly being used for this. However, questionnaires of this type tend to yield qualitative rather than quantitative responses and are not usually useful in a cost-benefit analysis. Rating scales can be used in questionnaires but tend to be research tools rather than workplace tools.

Questionnaires can be used in cost-benefit analysis if well thought out and the questions directed to elicit numerical data rather than opinions. For example; do not ask 'are there any problems with your work?' but more specifically '*how much time* do you lose each shift due to breakdowns?' or '*how much time* do you lose each shift due to lack of materials supply?'.

Preparing and administering questionnaires, though, is a specialist area and beyond the scope of this book.

3.10 Comparison with other companies

This is a source of information that is under-utilised. In this book we come from the viewpoint that a safe and healthy workplace leads to higher productivity. Thus it should be feasible to examine health indicators such as the injury rate, workers' compensation premium, or employee absence rate to determine if the enterprise is below the average for that industry. If these health and safety indicators are high it can show that all is not well, or at least that other enterprises have done better. However, there is a caveat. Statistics are only useful for large numbers of people: for small enterprises statistics may indicate that the enterprise is above or below an injury rate for the industry as a whole, but the data does not have any real meaning. For example, a small enterprise employing ten people may never have had an injury (it is below industry average for this factor) but this does not mean that there is no potential for injury.

A qualified comparison can be made between the productivity of one workplace and another or against the average for that sector or industry, but this requires general knowledge of the industry, knowledge which may be gleaned by management from experience, contacts within the industry, from industry (trade) newspapers, industry organisations and service clubs, sales people, and so on.

In the Introduction to this book we refer to the work of Ahonen (1998), who surveyed 340 enterprises using cost-benefit analysis and was able to use the data to compare and encourage the individual enterprises within the survey to increase productivity. Blewett and Shaw (1996) have used the concept of 'best practice' (identifying an enterprise with an outstanding record of, for example, safety or productivity) to encourage enterprises that have low productivity to reach for a higher level: 'if this enterprise can do it, why can't you?'. Valuable though such comparisons are, one must be careful when comparing data from different workplaces. Examples of the difficulties, as well as the merit, that may be encountered in interpretation of the data are given in Chapter 5D.

One useful method to enable management to 'see the wood for the trees' is to ask somebody from outside (an ergonomist or a colleague from another workplace) to come and have a look around. The person does not have to come from the same industry, it may even be better if he or she does not, but is simply a person who can ask the question – 'why is . . . ?'. The answer may be that 'we have always done it that way', but the question serves to break the mindset and to start the process that eventually leads to an intervention.

In our experience, asking pertinent questions will elicit new thoughts on the part of the person being asked and that may be all that is required to open the door to a new idea or concept, or simply that there may be better ways. Often that is what any competent health and safety practitioner will do – 'I wonder why such and such is done that way?'.

This brings us back to the earlier concept that the process of asking relevant questions is critical. The end result of a cost-benefit analysis, the pay-back period and savings, is simply a method to convince management to implement the intervention, whether the intervention be in the field of ergonomics, safety, training or human resources.

All a matter of asking the pertinent questions

In a small metalworking company certain forged pieces needed to be hand-finished. However, there was a high reject rate of the hand-finished pieces and also a high injury rate to the arms and wrists of the workers finishing the pieces; it was clear that the two, poor quality of work and high injury rate, were linked.

Figure 3.1 illustrates the awkward and forceful hand, wrist and arm movements involved in hand-finishing. It is not surprising that, towards the end of the shift, tiredness led to decreased quality. Eventually the awkward and forceful arm and wrist movements led beyond tiredness and resulted in injury.

The person the company called in to assist them, although an ergonomist, decided to let them fix their own problems; after looking at the way they worked he just asked the question: 'why do you hold the work pieces still and revolve the arms and hands?'.

The question was sufficient to prompt the company engineer, whose wife was a potter, to say: 'my wife holds her hands still while the clay spins and she does not get pains to her arms or hands'.

Figure 3.1 A static posture, together with force for the left hand and a forceful turning motion of the right hand, led to upper limb injuries.

Figure 3.2 With the introduction of the potter's wheel to turn the piece, the hands stay still with little or no force required of either hand.

After the company installed the potters' wheels (Figure 3.2) there were no new cases of injury and there was an increase in quality with a dramatically reduced reject rate. Based on the increased quality of the product alone, the pay-back period was less than one month. The workers' compensation premiums were reduced by 60% in the next year, with a further 25% the year after.

3.11 Productivity in the service industry

It is often assumed that it is difficult to measure service work and, if this is the case, how are we to get any relevant data? A valid question but one that smacks of defeat before one has even started. Much of the data is readily present in the form of wages and salary, absence rates and so on, even though output may not be able to be measured in the same way that one can measure output in manufactory. In a manufacturing plant one can measure output in the form of solid objects but often there are similar measures in service work; for example, in a mail centre articles handled per person, or in a call centre calls per unit time.

Clearly such measures are not always appropriate; call frequency in a call centre can only be a measure of productivity if all calls are similar in time and nature (e.g. directory enquiry). But even so, when there is a large degree of variability then a measure can be the mean taking into account the standard deviation.

These are measures of quantity, but quality can also be measured. For example, in call centres operators are often monitored and their achieving or not a set of quality standards (key performance indicators) is measured regularly. This measurement of quality can be used to determine the amount of any extra training needed, which can be costed.

For data entry a measure can be keys struck per operator per unit of time (quantity measure) or accuracy of input (entry is usually double-checked; quality measure) but there are often limitations to obtaining such data as they may encroach on privacy or discrimination legislation. Measurement of team, rather than individual, output may overcome some of these privacy issues. To open this privacy or discrimination door, if present, will require trust and cooperation between the employees, employer and union as well as the investigator.

Productivity measures for an office worker – time at work

The following case describes the cost incurred by a financial firm when one of their workers, an accountant had periods of absence due to back pain.

This accountant enjoyed her work and did not consider her complaint to be serious but, as time went by, the frequency and duration of her absences increased. The specialised nature of her work meant that it was not practicable to employ a replacement on a casual basis so that the time lost through absence was a net loss to the employer. Time lost in a year was nearly 10% of the accountants working time.

To reduce the absences the employer decided on work station improvements and rehabilitation for the accountant. The office was redesigned to reduce the amount of stretching and twisting of her back and the occupational rehabilitation course taught her how to manage her back condition by developing skills to minimise her disablement at work and at home.

Subsequently, the accountant had no further absences due to back pain. The gain to the accountant was in freedom from path and the gain to the employer was nearly 10% of the accountant's working time.

Details of the case are given in Chapter 5G.4.

Another measure of output, especially in intellectual work, is to measure the time that one is at work. The assumption here is that, even though output varies day-by-day, over a long period (several months) it will average out; a measure of quantity. The intervention will be aimed at keeping people at work, rather than off work, and thus the measure of success for the intervention is increased time at work.

3.12 Individual variation

One criticism often cited of productivity measures for service workers is that they ignore differences in quality of work between people; that one worker performs his/her tasks more competently and productively than another worker. But that is also a criticism of manufacturing; one person can operate a machine more competently and productively than another – fewer rejects, less waste and so on. Everybody who has had a carpenter or plumber to their home knows about quality variation between trades people.

It is both a weakness and a strength of cost-benefit analysis that individual variation is ignored. It is a weakness in that individual efforts are assumed to be all the same, and a strength in that one does not use cost-benefit analysis to determine the suitability of individuals for particular tasks; that determination should be left for personnel management, not for financial management.

The concept in economics that people will react in the same manner given the same information was discussed in more detail in Chapter 2.

3.13 Was the intervention a success?

It is always good management practice to evaluate any changes that were made. It is just as well to bear in mind that all data used for future costs are only estimates based on the best knowledge available at the time: the future can only be a forecast.

Too often changes take place with no regard for verifying whether or not the changes have been beneficial. In many cases the instigator of a new system or practice does not want it to be known that he/she has put in place a failure and would rather not know whether the changes were good or bad; a safe but unimaginative approach and one that does not allow learning to take place. We tend to learn more from our mistakes than from our successes, even if mistakes do lead to loss of a good night's sleep.

The concept of a cost-benefit analysis is to predict what will happen and it is by going back after the intervention has taken place that one can see whether the prediction was correct or not: that is the learning process. It can give you confidence that the process is suitable and demonstrate to management the value of ergonomics interventions.

Using the traditional formal (scientific) logic, it cannot be said that good ergonomics will *always* lead to good economics; it would then take only one case to disprove the entire concept. Rather, it has been our experience that well thought-out ergonomics, or good health and safety practice, almost invariably leads to improved financial outcomes. We have found that in the majority of cases the pay-back period is less than six months.

There is no general rule of thumb that will say what is acceptable as an economic success in all industries. In some industries (the 'high tech' ones, for example) the investment has to be profitable in a very short time due to the pace of change in that industry;

in a more traditional industry the investment is only expected to give a return after some years. The financial return will have to match the expectations of the industry or enterprise for short- or long-term investment.

As the primary aim of the occupational health and safety practitioner or ergonomist is to improve working conditions, the use of economics to support ergonomics needs to be seen in this light. One must be aware that it is the economics that supports the ergonomics, and not the other way round.

3.14 References

Ahonen, G., 1998. The nation-wide programme for health & safety in SMEs in Finland. Economic evaluation and incentives for the company management. *Protection to Promotion. Occupational Health & Safety in Small-scale Enterprises. People and Work.* Research Report 25, Finnish Institute of Occupational Health, pp. 151–156.

Blewett, V. and Shaw, I., 1996. *Benchmarking Occupational Health & Safety* (Canberra: Australian Government Printing Services).

Deming, W.E., 1982. *Out of the Crisis* (Cambridge: Cambridge University Press), p. 11.

4 The Productivity Assessment Tool

4.1 What this chapter is about

The principles behind various economics analysis models are discussed in Chapter 2 and the data needed for a cost-benefit analysis in Chapter 3. This chapter discusses one assessment model, the Productivity Assessment Tool. A basic version of the program (productAbilityBasic) which uses the Productivity Assessment Tool is enclosed with this book as a CD. It will allow you to follow the calculations used in the case studies in Chapter 5 and to make analyses from your own workplace.

Section 4.2 of this chapter provides an overview of the Productivity Assessment Tool including an introduction to the terminology used. Section 4.3 provides instruction on running productAbilityBasic, and the last sections of the chapter discuss applications for the Productivity Assessment Tool and integrating the results into reports.

It should be stated from the outset that the basic program (productAbilityBasic) is not the complete program. The complete program (productAbility) allows you to have up to five employees or employee groups, each having different hours of work and different pay rates. For example, you can have full-time employees (38 hours per week; $600 per week), casual employees (20 hours per week; $20 per hour), casual employees (30 hours per week; $18 per hour), temporary employees (15 hours per week; $22 per hour), each group with different overtime and productivity levels.

In addition the complete program allows you to compare the initial state of the workplace with up to four potential interventions. To take the case of a warehouse, interventions designed to reduce back injuries could include: extra staff; another fork-lift truck; improved shelving; changes to packaging. The productAbility program allows you to make direct cost-benefit comparisons between these four potential interventions.

The basic program is rather more restricted as it allows only one type of employee with one intervention. However, it is a fully workable model and will allow you to make a cost-benefit analysis of your own workplace and to print the results. The 'Screen Assistance' is displayed alongside the data screens so that the concepts and assistance are there when needed. Additional help screens are also given by clicking on the Help menu or F1 and in 'Handy Hints'.

4.2 The basic principles of the Productivity Assessment Tool

4.2.1 A decision-making tool for managers

It always seems a reasonable adage that 'what can be measured can be acted upon'; frequently management does measure before acting, but not always. However, as the purpose of this book is to encourage the measurement of the costs and benefits of health and safety programs and human resource interventions in the workplace, we will assume that the adage is correct. It is certainly the case that management cannot act rationally or in the best interests of the enterprise if there is no measurement preceding the action or measurement of the results of that action. It is a pedagogic principle that 'one learns by doing', but unless you measure the results of the intervention there will be no learning.

So, the notion behind using a cost-benefit analysis is to allow management to assess the value of any particular action and to compare it with alternative actions. From occupational health and safety practitioners' point of view, there is nothing to be feared here. The pay-back period is frequently so short that it compares favourably with any alternative action or investment by management.

By using cost-benefit analysis within the workplace, the occupational health and safety practitioners, ergonomists and human resource personnel no longer are seen to be a cost to the enterprise; they are a part of the enterprise and as keen to increase profits as the next person. A catchy phrase is 'Good ergonomics is good economics'.

Perhaps in this age of 'down-sizing' or reduction in employee numbers it should be stressed that a company consists of its employees, not simply its machinery or computers. It is its employees that run the machinery which eventually produces the profits; without people there are no profits. Despite the view of the accounting procedures that wages are just an expense, wages can also be on the benefit side of the profit equation and the proper description of these costs demonstrates this.

Remember the three 'P's: *'People Produce Profits'*.

4.2.2 Overview of the Productivity Assessment Tool

The Productivity Assessment Tool has four parts:

- data concerning the employees includes the numbers of employees, their working time and wages, supervisory costs, overtime and productivity
- data concerning the workplace includes recruitment, insurance and other general overheads, waste and energy use
- the intervention. This is the costs, or estimated costs, of the test case(s) for the intervention
- the reports. Cost-benefit analysis calculations and reports regarding the differences (savings and pay-back) between the initial case and the test case(s) for the workplace and the employees.

Figure 4.1 The concept of the Productivity Assessment Tool.

	Initial Case Enter data on:	Test Case(s) Enter expected **changes** for:
Data concerning the employees	• productive hours • wage costs • overtime • reduced productivity	• productive hours • wage costs • overtime • reduced productivity
Data concerning the workplace	• recruitment • insurance • reduction in waste • energy use • other overheads	• recruitment • insurance • reduction in waste • energy use • other overheads
The intervention		costs, or estimated costs, for the intervention
The reports	Cost-benefit analysis calculations and reports of the workplace and the employees	

The basic version (productAbilityBasic) is restricted to one test case and one employee or employee group. The complete program (productAbility) allows up to four interventions (test cases) and up to five employees or employee groups in each case, groups being determined by similarity in work tasks, hours worked and wages.

4.2.3 Concept of productive time

If you are producing solid materials (nuts and bolts, textiles, pencils, and so on) then machine or materials productivity may be measured in terms of output per unit of time (pencils per hour, and so on). Some types of worker productivity can also be measured as the output that a worker makes in a unit of time, but often output is not quantifiable in this way and needs to be supplemented by other measures. One measure, and in some cases the only measure, of productivity is the ratio between the time paid for by the employer and the time the employee spends actively working (the productive hours).

Productive hours are defined as the total hours paid for by the employer *less* hours not actively producing. The Productivity Assessment Tool calculates productive hours over a one year period.

These 'non-productive' hours, which are paid for by the employer, include:

- injury (workplace) absence
- illness absences
- training
- vacation and statutory holidays
- other absences, e.g. for maternity leave, military service.

Productive time (hours on a yearly basis) is calculated by the program from data you have entered in the 'Employee Details' screen, which is shown in the Employee Summary report.

The call centre industry, for example, regards daily productive time as a measure of *productivity* and mostly takes great pains to ensure that, each day, the operators are plugged into the telephone system as much as possible. Eighty per cent of the time plugged into the telephone system is assumed to be more productive than 75%.

However, it is not sufficient to look at each *day* at work; it is necessary to look at the total picture over a longer period, say a *year*. Unnecessary days off work due to a poor working environment will lead to a loss in productivity which may not be measured by the daily productive time.

The so-called knowledge industry (architects, scientists, physicians, authors, etc.) is an example where quality is the crucial parameter which cannot be measured by output, or can only be measured with difficulty, but in many cases an approximation to productivity may be measured by productive time. Within reasonable limits, professional people will produce more if they are at work for a longer period over a year than for a shorter period, and for many the *total productive time* may be the only productivity parameter that is easily measurable.

Clearly, the drawback with the concept of productive time is that it takes no account of quality. We make no apologies for this when presenting a simple analytical tool; attempts have been made to quantify quality but they tend to be difficult to use. In measuring the productivity of a law court it is hardly sufficient to look at the productivity of a judge (measurement being the number of cases each year) unless one looks at the quality of each judgement (fairness and agreement with the law). This measurement has been attempted but required a multidimensional analysis with a great deal of input from research workers and is not a tool that can be generally used. Hence, in many cases productive time may be the only reasonable measure and other quality differences can only be incorporated descriptively into reports.

4.2.4 *Measuring productivity losses from the 'ideal state'*

As noted in the previous section, the productive time is measured over a yearly period and is calculated from data entered in the 'Employee Details' screen.

The term 'Reduced Productivity' is used to record all reductions in output below the maximum possible: the 'ideal state'. In any workplace there is always some knowledge of the 'ideal state'; if all the machinery worked perfectly and if only people did not get tired or wish to talk to their colleagues or even go to the toilet, what we could achieve! The Productivity Assessment Tool assumes that this figure is known and asks 'how close are you to this 100% ideal state?' and thus 'what is your reduction in productivity?'.

In some industries this 'ideal state' can be known accurately. In manufacturing, the number of items that a particular machine can make in a certain time period is taken to be 100% and anything less than this is converted to a percentage reduction; e.g. if 1,000 is the maximum number of items that can be made on a particular machine in a full shift of eight hours and the actual production is 900, then the productivity is 90% and the reduced productivity is 10%. The 10% loss may be due to breakdown of the machinery, a poor working environment or any of numerous variables. In the Reduced Productivity data screen the 10% loss would be recorded in the box(es) that most suits the reasons for the loss – say, 7% due to the equipment and 3% due to low skills of the operators.

In other industries the 'ideal state' is more difficult to measure and to quantify, and it is for this reason that the Productivity Assessment Tool allows you to enter reduced productivity in a number of ways:

- as a percentage of the reduction from the maximum output
- as time lost from production on a shift or short-time basis due to low skills, inappropriate hand tools and so on.

The daily time that call centre operators are plugged into the telephone system can be used as an example. The 'ideal state' can be determined starting with the total available paid time per shift and taking away the time required for breaks, supervisor and team meetings, training and any 'paperwork'. If the actual time plugged into the telephones is less than this it will be a loss of productivity and thus a measure of reduced productivity. In the Reduced Productivity data screen, if you assume that the maximum reasonable time (the 'ideal state') to be plugged into the telephone system is 80% of the working day (i.e. 384 minutes out of an eight hour shift) and only 75% is actually achieved, then the reduced productivity is $((80 - 75) \div 80 \times 100) = 6.3\%$.

One can make a similar determination about other service industries such as retail shop staff and word processor operators/typists/secretaries.

> **Reduced productivity**
>
> Reduced productivity is the *loss* of production due to measurable factors that include:
>
> - low skills of the employees
> - unsuitable hand tools
> - machinery not maintained sufficiently
> - unsuitable machinery
> - physically strenuous work with too few rest breaks
> - unsuitable environmental conditions (lighting, temperature, noise)
> - poor work-station layout.
>
> The reasons for these factors should lead to interventions addressing and correcting these factors.

4.2.5 Allocating indirect costs

Some of the measures used in the Productivity Assessment Tool require an allocation, or a proportion, of the indirect employment costs. This is often a percentage used by companies routinely for head office costs, say 5% added to the wage bill. Allocation can also be made for the costs of supervision within a company; for example if five employees share a supervisor who spends 50% of his/her time with them, or on matters connected with them, then the employee group has a 50% share of the supervisor's wages to accommodate this. Half of the supervisor's direct wages are entered either as employee supervisory costs in the Employee Costs screen or in the Allocated Costs screen (do not double-count by entering the data in both places).

The same is true for the personnel/human resources department. If there are 100 people in the factory then 1% of the wages for the personnel/human resources department is allocated to each employee. This figure can be entered as Employee Administrative Costs in the Employee Costs screen.

With indirect costs, most frequently the direct wages of those employees are sufficient. It comes down to determining how much data you really need and how much effort you need expend to get data that may not add substantially to your arguments. Do you need to add all the indirect costs?

4.2.6 Reduction in productivity and the role of interventions

The points above are intended to focus the mind on the workplace and to ask the question 'are we achieving the optimal reasonable productivity?' Measurement will tell us 'yes' or 'no' and if 'no' will lead us to the next question 'why?', followed by 'what can be done?'

The answer to 'why?' may be an engineering solution outside the scope of this book but frequently the solution will concern people. The reasons that employees give lower than optimum productivity may be innumerable: too much stress, too few breaks, poor supervision, poor lighting, uncomfortable seating, low pay, low status and so on. Some of these are within the scope of ergonomists, occupational health and safety practitioners and human resource personnel and some beyond their scope, but do not dismiss any aspect too lightly. It is true that occupational health and safety practitioners are not usually in a position to influence pay but it is possible to influence supervision, training and work methods as well as other and more obvious aspects of poor lighting, ventilation and uncomfortable seating.

The intervention (test case) is the change proposed to ensure better and safer working conditions; measuring reduction in productivity due to poor working conditions is one means of showing the benefit of the intervention.

4.2.7 The pay-back period

The economic parameter chosen for the Productivity Assessment Tool is that of the pay-back period which relates to the improvement in costs due to the intervention; it is simply the time taken to pay back for the intervention.

The formula is:

$$\frac{\text{Costs for improvements (the intervention)}}{\text{Benefits or savings}}$$

Both costs and benefits are measured over a one-year period.

This is a simple measure and does not take into account the cost of borrowing money or other long-term economic factors. In our experiences with ergonomics interventions this is not a drawback, as the pay-back period is frequently just a few months: using measures to calculate the net present value (NPV) or return on investment (ROI) add little to an intervention with a short pay-back period but may be required if the pay-back period is long. This is dealt with in more detail in Chapter 2.6.2.

4.3 Running the Productivity Assessment Tool

For technical details and loading the productAbilityBasic program, please refer to Chapter 8. This present section describes the use of both the basic version (productAbilityBasic) and the complete program (productAbility) and will serve as the software manual. The steps will be easier to follow if you have the software, productAbilityBasic, open on your computer.

4.3.1 Overview

The program opens with the 'Open productAbilityBasic files' screen. Select (click) 'Cancel' to start a case of your own or select a case study from Chapters 5 or 6 (File/Examples). The 'Handy Hints' screen appears when you start a case; more information is given in the 'Help' screen which can be displayed by going to Help/Help or F1.

The data entry screens consist of a main menu and icons for selecting files, print and so on. Below this is the section for selecting the data entry screens, reports and a box for selecting the initial case or test case(s). Clicking with the left mouse button opens the selected screen (Figure 4.2).

Case\Full-Time Employees

View\Set Currency

File Edit Case Print View Help

Select a table

▼ editing Initial Case

◆ 0. Home

1 Employee Details
2 Employee Costs
3 Reduced Productivity

4 Allocated Costs
5 Recruitment Costs

7 Workplace Summary
6 Employee Summary
8 Workplace Report

Workplace Details

Investigator : MBO

Company : T&F

Workplace : London office

Welcome to productAbility, the software for the Productivity Assessment Tool

This your home page where you enter your details and those of the workplace where the intervention is to take place. Working details are given in the 'Handy Hints' and in the 'Help Screens' (Click on 'Help' or F1)

You may enter the hours for a full time-employee (click on 'Case') and the currency you wish to use (click on 'View\Set Currency').

At the foot of each data entry screen you may add your notes.

You are now ready to enter the data for the Initial Case. Click on Employee Details.

This space is for your notes

Figure 4.2 productAbilityBasic: the first data screen.

The left-hand side of the data entry screens is where information is entered. On the right side of the screens is the Screen Assistance for that particular data screen. At the foot of the screen is a place for you to type in your notes; the amount of space for notes can be expanded or contracted by placing the cursor over the top border of the notes box, clicking and dragging.

If you are using the Productivity Assessment Tool for your own workplace, we suggest you firstly familiarise yourself with some of the concepts of the program through the case studies in Chapter 5 or 6 (File/Examples). You may find that one of the case studies has similarities to the intervention you are proposing and can help you to identify data you need.

Enter the data for the Initial Case (the Initial Case is the present situation or as it was prior to the changes or intervention). Especially for your first trial run it is best to limit the workplace to as few employees as possible.

When you make a test case all the data you have entered for the initial case will be transferred automatically to the test case; you then only have to change that data which the intervention will affect, saving time and reducing the possibility of data entry errors.

The results of the cost-benefit analysis will be shown in the two summary reports: 'Workplace Summary' and 'Employee Summary'. The Workplace Report gives the details of the data and notes that you have entered and is useful for checking that you have entered data correctly.

A word of caution; as with all computer programs – save your work regularly.

4.3.2 Home screen

The Home screen is where you enter data about the workplace. After you have entered the workplace description on the Home screen go to the main menu View/Set Currency and ensure that the symbols used ($£$, $, etc.) are the ones you prefer or enter your own. You should also enter full-time employee hours by going to Case/Hours per Full-Time Employee. The default is 40 hours per week.

4.3.3 Employee details screens

Employee Details data screen

In the basic version (productAbilityBasic) only one employee or employee group is allowed but up to five employee or employee groups can be entered in the complete program (productAbility) (see Figure 4.3).

Name the employee group or category and enter the number of employees and the number of *paid* hours worked by each employee in this category (which may be the same or different from the hours per full-time employee). In the complete program, for any one category the workers should work approximately the same number of paid hours per year.

Enter paid time lost (unproductive time; total hours per year; section 4.2.3). You may do this by total hours per year for the whole employee category or for each employee by clicking on one or other of the radio buttons.

Figure 4.3 Overview of the Employee Details data screen.

Employee 1	In the basic version of the Productivity Assessment Tool only one employee or employee group can be selected. In the complete version up to five groups can be used. The name of the employee/group can be changed in this screen.
Number of employees	In the selected workplace only.
Hours per week	The paid work period can be selected (day, week, etc.) but is calculated in the program as hours worked/year and as a proportion of a full-time employee.
Enter figures for this employee category	Select this if you are entering figures for paid absence for the entire group.
Enter figures for each employee	Select this if you are entering figures for paid absence for each employee.
Illness absence	Absences for illnesses and paid for by the employer (hours/year)
Injury absence	Absences due to injury at work (hours/year)
Training	Training during working time but do not include initial training for new employees (hours/year)
Planned absences	Vacation and public holidays paid for by the employer (hours/year)
Other	Maternity leave, military duty, etc. (hours/year)

Data should be entered for the following categories that are relevant to your intervention.

- Illness Absence refers to short-term absence.
- Injury Absence refers to absence due to compensatable workplace injury.
- Training refers to absences from production or service for skills/competence training. Do not add training for new employees here; such costs are added to the Recruitment Costs data screen.
- Planned Absences refers to absences for vacation, statutory holidays and similar absences.

Use the Tab key to move from one entry point to the next.
For more details see the Screen Assistance and the 'Help' screens (F1).

Employee Costs data screen

This is the gross wage cost for each employee or employee group and does not include overtime. As well as the wages or salary paid directly to the employees the obligatory costs should be added as well as the allocated costs (proportion) of supervision and management.

Under the screen headings Employee Administrative Costs and Employee Supervisory Costs you can choose the cost for each employee or the total of all employees in the selected employee group by clicking on the appropriate radio button.

Employee administrative costs include:

- administrative costs
- pension (superannuation) fund
- workers' compensation premium (unless added to the 'Allocated Costs' screen)
- taxes paid by the employer on wages
- personnel/human resources department
- medical and direct injury costs
- head office allocation.

The employee supervisory costs include direct supervision (a manager/foreman/supervisor in the workplace) and indirect supervision (middle/senior management).

The administrative and supervisory costs may need allocating on a proportionate basis, as explained in section 4.2.5.

If it is more convenient, some of these costs can be entered in the Allocated Costs data entry screen but be careful not to enter the same information twice.

The overtime setting is for the overtime worked over the whole year by the employees in the employee group(s). You can enter either the total annual cost for the selected employee group or the rate per hour and total number of hours annually. Please note that the figures are for the *total annual* amount.

In the test cases you can either add the annual cost or the annual hours expected from the ergonomics intervention or use the slides and observe the calculations shown simultaneously at the foot of the screen.

Reduced Productivity data screen

In some ways this is the heart of the Productivity Assessment Tool. Up to now the figures obtained mostly are data from the accounting or pay system and can be taken to be relatively accurate; the Reduced Productivity data screen may use much softer figures (see Figure 4.4).

Figure 4.4 Overview of the Reduced Productivity data screen.

Reason for reduced productivity	Typical interventions that would lead from these factors
Low skill	Improve or increase training
Hand tools	Replace or improve hand tools with a design better suited to the work tasks
Capital	Improve maintenance, replace machinery. Do not double-count by adding the same data to the 'Allocated Costs' data entry screen
Other	Factors not covered above

Reduced productivity (%) and cost per year for the employee group based on employee costs

Reduced productivity refers to losses from an 'ideal state' and you can use either actual figures (number of items produced per hour) or the time spent on particular functions, but both reduced to a percentage. The concepts have been explained more fully in section 4.2.4.

It is important that the reasons for the losses are identified as this will guide the selection of the best or most suitable intervention. As an aid to this selection the data screen indicates three broad areas: low skills, hand tools and capital equipment, as well as 'other'. For example, the reduction in object output in a manufacturing plant may be due to poor skills on the part of the operators and in call centres due to waiting for calls to come through. Be sure to add notes to the foot of the screen to explain the reasons for your determinations.

4.3.4 Employment overhead screens

The previous three data entry screens relate specifically to the employee or employee groups chosen. The next two screens relate to the workplace as a whole and may not specifically refer to the individual employee or employee group; in these cases costs need to be allocated proportionately. This is especially the case with the complete program, where there can be more than one employee group and costs are allocated across several employee groups.

Allocated Costs data screen

In many workplaces, particularly larger ones, there are overheads that should be allocated to the cost of employment and a proportion of these overhead costs related back to the employee groups (see 4.2.5). For example, if the enterprise employs a total of 100 people and ten of these are 'non-productive' (administration, pay clerk and so on) then the cost of these ten should be distributed (allocated) among the other 90 people. This is also the case for the workers' compensation premium and other overheads that you may not have entered in the Employee Costs screens.

If using the complete program for several employee groups, rather than the basic version where only one employee group can be used, it may be simpler to add the overheads to the Allocated Costs data screen rather than to each of the Employee Costs screens. Be careful not to add the same data to both the Employee Costs and the Allocated Costs data screens.

That part of the insurance (workers compensation) premium relating to the employee groups should be entered here. If the intervention is to reduce injuries then one can expect that there will be a later reduction in insurance premiums. Strictly speaking, as this will not occur in the year of the intervention the reduction in insurance premiums in later years should not be entered into the test cases. However, as described in the hospitality industry case (Chapter 5B) this was an important cost reduction factor and its addition to the cost-benefit analysis in the year of the intervention, rather than in the following year, is warranted if noted as such.

In many cases the intervention will not affect overheads so there will be no difference in these costs between the initial case and the test cases and thus no apparent effect on the pay-back period. However, using this data will make a difference to the demonstrated cost of employment and to the amount that management might think it reasonable to spend on an ergonomics intervention.

Equipment Running Costs refer to the machinery, computers, energy use etc. that are used at the workplace and that may be affected by the intervention, or that you wish to add as a significant cost factor. For example, maintenance costs (see Chapter 5E.3) may be a significant cost factor that may be affected by the intervention and may be a factor considered important in the overall costing of the product or service. This data screen does *not* refer to the intervention itself; the intervention costs data are added later (see 'Cost of the intervention' below).

Recruitment costs

These costs relate to the employment of new employees and the loss of skills when employees leave. They may be regarded as 'soft' figures but assume importance if the intervention affects the turnover or retention of employees.

Employment of new employees indicates the need for some training, with the amount of time depending on the type of work and the previous experience of the new employees. This is a figure that is easily obtained if the training is classroom-based but, even if the training is sitting alongside an experienced employee at the work station, there is still a cost for wages for the new employee and possibly a reduced production by the experienced employee. These costs need to be estimated if significant to the intervention.

Loss of an experienced employee is often a cost hidden or simply not realised. For example, it may take a senior manager a year to understand a new environment and you can see this either as a training cost or a loss of the previous manager's knowledge and experience. On the other hand, a nurse may require little additional training as he/she will be trained in a technology which is similar from hospital to hospital within the same system.

As a rough guide, one can assume a loss of 50% of production during training of a new employee if no better figure is available.

4.3.5 Entering your intervention data

Although data can be changed at any time, once you have entered as much as you reasonably can for the initial case, you can now run your test or intervention case(s).

All the data entered so far will be reproduced in your test case(s). It is not necessary to enter all data requested or even to use all the data sheets; the minimum data needed to run the program is the number of employees and their working time and pay; all other data will increase the accuracy of the result. As noted above: 20% of the time is spent collecting 80% of the data, but then to spend 80% of the time to collect the extra 20% is rarely worthwhile.

Click on 'Editing Initial Case' in the top left section of the screen and click on 'add test case'. The box will now read 'editing new case' and you can choose a suitable name in the Allocated Costs data entry screen.

Cost of the intervention

The Allocated Costs screen for your test case(s) is where you add the cost of the intervention. The Service Costs include management and consultants' time and the Capital Costs are for machinery and other goods required for the intervention.

The effect of the intervention, either estimated or measured, is entered in the 'Employee Details' and subsequent data entry screens.

The resultant calculations can be seen in the three report screens: Workplace Summary, Employee Summary and Workplace Report. These report screens are discussed in sections 4.1 and 4.5.

The calculations are based on the sum of the total costs for the initial case less the total costs for the test case(s). The resultant calculation will be the savings per year due to the intervention. The pay-back period is calculated as described in section 4.2.7.

4.4 Applications of the Productivity Assessment Tool

4.4.1 *Productivity measurement (simplest use of the Productivity Assessment Tool)*

When you enter the relevant starting situation (initial case) and the corresponding data for a projected situation (test case or intervention), the Productivity Assessment Tool will calculate the pay-back period and the yearly savings. As a general rule, ergonomics interventions have a short pay-back period.

Workplace Summary report

Productivity measurement is the simplest and most frequent use of the Productivity Assessment Tool. The results of this calculation are shown in the Workplace Summary report screen and give the major information needed by management.

This summary screen shows the number of equivalent full-time employees (EFT) in the chosen work station, the total cost of employment and the cost of the intervention. The savings due to the intervention are given along with the pay-back period.

The number of equivalent full-time employees is a straightforward calculation based on the information given in the Employee Details data screens.

The total cost is based on the summation of the five cost screens (employee data and workplace data) and this is used in the subsequent calculations for the savings due to the intervention. Once the total costs have been calculated the difference between the initial case and the test case(s) is the savings per year due to the intervention.

The pay-back period is the cost of the intervention divided by the savings made due to the intervention. This is reported in months as, in our experience, most ergonomics interventions occur in less than six months; after the pay-back period the savings are all 'profit'.

Employee Summary report

This gives details of the wages and the cost per productive hour of the employees by individuals and by the employee groups or categories. It is more detailed than the Workplace Summary report.

Workplace report

This is a detailed report of the data you have entered. It is designed for the investigator to check for accuracy of the data that has been entered.

4.4.2 *Of several alternatives, determining which intervention will be the most cost-effective*

The Productivity Assessment Tool allows comparison between alternative means for improving working conditions. If, in any workplace, there are alternative means for improvement then the program allows a direct cost-benefit comparison between these alternative means. The basic version allows only one test case and is not suitable for this type of comparison.

It may be initially thought that a cost-benefit analysis will not indicate the best ergonomics solution but only the lowest cost solution. However, if the cost of injuries is added as well as productivity factors, the most cost-effective solution will usually be the most suitable ergonomics solution too.

For example, if the project is to reduce back injuries in a warehouse and the alternative means are increased staff, better shelving, another fork-lift truck or automation, by putting in the projected costs and benefits for each a comparison can be made. Better shelving may look to be the cheapest solution but, as the data you enter will show the injury costs as well as productivity, it may not be the most effective solution. Although a fork-lift truck may cost more initially, a higher reduction in injuries and better increases in productivity may well indicate that it is more cost-effective and ergonomically more sound than better shelving.

4.4.3 *Sensitivity analysis*

If one is to make an ergonomics intervention it is just as well to identify those parameters that, by their alteration, will lead to the greatest cost-benefit. That is, if you have to 'sell' the intervention by means of it being cost-effective then you will need to know that the intervention will tackle the most sensitive cost parameters.

The Productivity Assessment Tool can be used for a sensitivity analysis by entering the data in the initial case and then in the test case. You then take a parameter such as overtime or reduced productivity and change the selected parameter in the test case by a reasonable amount. For example, if you have entered in the initial case, say, 10% reduced productivity, then in the test case by using the slide you will see at the foot of the data table the change in yearly productivity cost. If you alter the Employee Costs data screen the changes will be shown in the employment costs in the Employee Summary screen.

By going to the Workplace Summary report you will see the pay-back and the yearly savings results and can make an appropriate judgement.

4.4.4 *Rehabilitation*

In many legislations there is compulsory rehabilitation for a worker injured at work, although there is usually some leeway about how much effort and cost the employer need apply. Rehabilitation is often effective and a cost-benefit analysis can be used to give some indication of this (see the case studies Chapters 5G.3 and 5G.4).

In any event, rehabilitation of employees may make both supervisors and workers re-evaluate the workplace in terms of improved occupational health and safety. The cases give an indication of the value of a cost-benefit analysis in rehabilitation of injured workers.

4.5 Reporting and decision-making

The program has three reports, of which the 'Workplace Summary' is the most important for management and decision-making. This report gives the total employment cost of the workplace chosen (limited only by the amount of information entered), the savings due to the intervention and the pay-back period.

The 'Employee Summary' shows the calculated data for each employee and each category of employee including the wages, productive times, reduced productivity and overtime costs and employment costs.

The 'Workplace Report' is a long report as it shows the data entered and is useful for ensuring that the correct data has been entered.

How much of the program's reports you use will depend on the information expected by those who hold the purse strings. It may be sufficient for management (or your client if you are an outside consultant) to know that you have worked through the costing and they only need see the end result; basically the Workplace Summary. On the other hand, you may need to show the details of the assumptions you have used and variations to the outcome (savings and pay-back) if the assumptions are changed.

In the hospitality industry case study (see Chapter 5B) the ergonomics consultants were asked to investigate the reasons for the increase in workers' compensation insurance premiums and, after investigations, two reports were prepared. The first report defined why injuries were occurring and what could be done about them; that is, a straightforward ergonomics report that looked at the cause of the injuries and the prevention of further injuries. This report was used by management and employees, through their health and safety committee, as a basis for training and improving work methods and materials.

A second report was produced for management that showed the expected financial cost and returns (the cost-benefit analysis) of the project, using cautious assumptions that the number of days lost through injuries would be halved due to the intervention and the workers' compensation insurance premiums (the experience penalty portion due to the cost of the injuries) would, similarly, be halved. In actual fact, in the year after the workplace changes were introduced the injury absence rate had dropped far more substantially than anticipated, with only one major injury and several minor ones, and the insurance premium (experience penalty) had dropped by over 80%.

So, in deriving your cost-benefit analysis there are several points that you need to note and questions that you need to ask:

- define the work area and the people who work there
- obtain information on wages, absences, supervision and other costs
- can or will the productivity in the work area be improved?
- how will the intervention affect the people and the work area?
- what assumptions have you made regarding the effects of the intervention?

For the above you will need to work closely with the supervisors and employees in the work area and have the agreement of management on the cost parameters you choose. You should look on the method for deriving data and development of the analysis as one of iteration.

When preparing your report you will need to know:

- what level of information or detail management requires
- how different costing and assumptions will affect the outcome of the analysis.

You can use up to four test cases in the complete version of the Productivity Assessment Tool to show how different interventions will affect the outcome or how different costs and cost variations affect the outcome.

If you are presenting your report to management, a simple but effective method to illustrate the effect of different costs and cost variations proposed by management is to run a live demonstration of the Productivity Assessment Tool. In the test case(s) you can use the slide in the Reduced Productivity data screen, for example, to show how changes to productivity affect costs, savings and pay-back.

5 Case studies

Contents

5.1 The argument for and against better working conditions

Since the end of the Second World War many countries of the developed world have made great improvements in occupational health and safety. In many countries Robens, or similar, legislation has been introduced with the aim of creating safer workplaces; the employer was responsible for a safe workplace and it was no longer enough for the employer to say to a worker 'be careful' if the conditions were unsafe.

Such legislation led to an expectation among practitioners in occupational health and safety that it would be easier than previously to achieve safe workplaces. Unfortunately, the past decade has rather shattered that illusion. Injury, rather than the prevention of injury, is still a major driving force in making working conditions safer; practitioners may be called upon to make workplaces safer but they are more likely to be called on to pick up the pieces after an injury. On the whole, there has been little progress from *reaction* to *prevention*. Why is this so?

Although we had such high hopes for an improvement in working conditions we have been overtaken by events led by the increasingly intrusive effect of the market economy or economic rationalism. In most countries of the world the overriding economic paradigm is now the *market economy*. The command or communist economy has given way to capitalism and the welfare/socialist/capitalist states have veered further to the extremes of capitalism, less modified by government regulation and enforcement. Economics, rather than social and humanistic values, has set governmental and international policies.

Robens-style legislation

The Robens style of legislation, or legislation based on similar principles, has been adopted and adapted in many other countries. The concept was to simplify the law and to acknowledge that there must be a degree of responsibility from all parties, as described in the following extract.

> The key mechanism for achieving [safety in the workplace] was replacing much of the maze of existing regulation with a single overarching law, an enabling Act that established broad standards, decision-making structures and procedures rather than directly specifying the detailed standards that might apply to particular workplace, machine or work process. The new law would contain new general duty provisions that imposed broad-based standards on employers, designers, manufacturers, workers and other parties. These were articulated to a more systematic set of specification standards, performance standards or compliance guides expressed either in terms of codes or regulations made under the legislation. Significant avenues for employer and union/worker involvement also underpinned the approach, notably tripartite standard-setting bodies, bipartite workplace committees and employee health and safety representatives. At the same time, Robens urged a number of changes to enforcement including increased inspectoral powers and new sanctions (most notable, improvement and prohibition notices)
>
> (Bohle and Quinlan, 2000).

The trend towards the market economy has carried with it short-term goals in enterprise management, and the daily share price has become an overriding concern of management. No longer is the long-term viability of an organisation necessarily the main aim of management; long-term viability has to 'compete' with short-term aims and profits so that the share price is maximised. Some companies even reward their senior executives with bonuses in company shares – higher share prices means higher income when they sell their shares, a veritable invitation to short-term vision and profits. There is a trickle-down effect from the larger enterprises to the smaller ones so that the psychology of the whole economy becomes geared more and more to the short term.

Accompanying the trend towards the market economy is the fellow traveller, globalisation. Globalisation brought with it the concept, among others, that if the developing world has poor working conditions as well as low pay then, to compete on the price of goods and services, the developed nations must also reduce working conditions and pay. That is, the workers living in the developed countries should give up their hard-won better working conditions and reduce their conditions, including safety, to the lowest common denominator. Perhaps this is the meaning of the euphemism 'the level playing field'; it would certainly seem easier to bring good working conditions *down* rather than raise poor working conditions *up*. Lowering working conditions in the developed countries ignores ethics and compassion and the potential to improve working conditions in the developing countries.

As well as globalisation there have been other structural changes in employment conditions. There has been a continuing movement from the manufacturing sector to the service sector, bringing with it more casual and part-time workers (precarious workers) and less security of employment. Unions have become weaker in the face of a declining membership due partly to this change in employment style. Unemployment is now endemic in most countries and is being accepted as such: unemployed people, being mostly the less educated and poorer members of society, have less economic and political power.

Whether or not unemployment is a necessary condition or a consequence of capitalism is an argument that has a long history, but unemployment certainly seems to play into the hands of the multinational companies. With a freeing of capital flow in the 1980s they can take their business wherever labour wages are lowest, with unemployment having the effect of reducing wages. The reduction of power of the labour movement (national unions and International Labour Organisation) and the increasing power of the 'market' (through not only the multinational companies but also the various international organisations, e.g. World Bank, International Monetary Fund, World Trade Organisation) have had a continuing deleterious effect on employment. As Professor Jean Pyle (2002) points out regarding women's employment in the developing countries, there is now a 'global assembly line' (clothing, electronics, toys, shoes and sporting goods) with corporate strategies typically having an adverse effect on women and their health but, alongside this, women are being pushed into the informal sector (service industries including sex work and domestic labour) just to survive – poverty levels rise.

Cost-cutting is a method used in private enterprises, usually to increase short-term profits but often as a euphemism for putting people out of work. Governments followed this cost-cutting vogue, and no longer is full employment seen to be the goal and responsibility of governments. There has been a gradual modification downward of these responsibilities; cost-cutting means a reduction of services and employment, no matter what euphemisms are used. Publicly owned utilities (electricity, water, transport, communication, etc.) for ideological market economy reasons (or 'economic rationalism') have been sold off at knock-down prices to private enterprise; ostensibly the selling price is to ensure that private capital is raised for the utility but more frequently it is to balance a government budget in difficulties. With the remaining services governments pretend they are corporate companies rather than 'servants of the people' and abdicate their responsibility of providing services to the communities under the banners of cutting costs and/or increasing efficiency and/or 'user pays'.

Legislation is in force in most places to protect worker safety and recently some countries have even strengthened the existing laws. As the inspectorate becomes smaller, due to cost-cutting, the laws, not being enforced, become more and more nominal: thus, there is excellent legislation but less enforcement. It is common experience that the command 'Thou shalt not steal' needs reinforcement from an active police force and, in a similar fashion, occupational health and safety laws require an active inspectorate.

In any event, the majority of people are employed in small and medium-sized businesses and the chances of an inspector calling in the absence of an injury incident has always been very small; inspectors usually keep to the larger enterprises. We have seen calculations that suggest that a small enterprise can expect an inspector's visit once in 20 years, which is a considerably longer period than most small enterprises are in business!

Thus occupational health and safety is placed in a difficult situation as there are two competing goals. The first is that occupational health and safety programmes are often seen to be long-term rather than short-term and have suffered because of this psychological dominance of short-term market economics, and the second is the lack of enforcement of existing regulations. Neither self-regulation, due to a smaller inspectorate, nor short-term goals augur well for health and safety in the workplace.

The national occupational health and safety research institutes of the developed countries have also been caught in this governmental cost-cutting trend. It is of little use having the scientists and physicians saying one thing (poor working conditions are bad for health) and the industrialists and politicians saying another (poor working conditions are the only way we can 'compete'). It is an unfortunate fact that physicians and scientists have less power than industrialists and politicians, so the research institutes suffer. A reduction in the budgets and effectiveness of research institutes has been the general trend over the past decade or so although, fortunately, there are some exceptions.

The changes in employment status have been accompanied by a changing requirement in training and education. Employers are reducing the general education or training for their workers (apprenticeships, scholarships) and, conversely, increasing numbers of people are going to universities and technical colleges; more than previously workers come to the job complete with skills. The training that employers now give is largely for *enterprise-specific* skills leading towards a bipolar employment system with a core of skilled workers (tertiary educated) at one end and a majority of unskilled and/or enterprise-trained at the other.

On the other side of the coin there is a gathering momentum against the excesses of capitalism, with a concerted movement to draw the attention of the public in the developed countries to the working conditions of peoples in developing countries. For instance, Nike, a manufacturer of sports clothing, has had considerable bad publicity due to the conditions of the workers in Asia that were manufacturing its clothes and shoes and this bad publicity, one hopes, has made the company modify the working conditions of its employees. The condition of children, in particular, in carpet manufacture in Asia and the virtual or even actual slave conditions for cocoa pickers in West Africa are other examples where buyers, the consumers of the developed countries, can, and perhaps do, exert pressure to improve conditions for these unfortunate people.

Consumer-led public movements are an essential part of progress towards improving working conditions in developing countries, and this is a more noble aim than bringing downward the working conditions in developed countries, all of which brings us to the purpose of this chapter. This purpose is to present case studies to support our notion that good working conditions are more productive than poor working conditions.

5.2 Overview of the case studies

In the first edition of this book most of the case studies came from the manufacturing industry. Unfortunately, manufacturing industries still give rise to serious injuries and the theme that most of these injuries are preventable by the application of ergonomics and other safety measures still holds true. However, there has been a drift in employment away from manufacture to service industries, partly due to:

- the continuing improvement in efficiency in manufacture, needing fewer people
- the advent of new technology
- an increasingly affluent population needing services more than goods.

In this chapter we use both service and manufacturing industries to illustrate the interaction between ergonomics and economics – that good ergonomics leads to good economics. The cost-benefit data and calculations from many of the case studies are reproduced in the CD that accompanies this book – the basic version of the Productivity Assessment Tool. The software files for the case studies are noted in the tables in this chapter.

We suggest that you also use these case studies as a starting point to assist you to understand the use of cost-benefit analysis and determine the links between injury prevention and productivity improvements. This will lead you to develop your own studies as an aid to your interventions whether they be in the fields of ergonomics, hygiene, training, occupational health, safety or human resources.

Chapter 4 is the 'manual' for the software for the Productivity Assessment Tool and, as well as describing the practical use of the Tool, describes the terms and concepts used. As you will note from reading Chapter 4, the terms used in the Productivity Assessment Tool are readily obtainable from the 'Help' screens and also in the Glossary.

Chapter 3 describes how and where to obtain pertinent data for your analysis and, most importantly, the degree of (in)accuracy needed for an economic or financial analysis, and Chapter 2 discusses some of the background concepts in economics.

5.3 References

Bohle, P. and Quinlan, M., 2000. *Managing Occupational Health & Safety*, 2nd edn (Melbourne: Macmillan), p. 264.

Pyle, J., 2002. Keynote address to the *Women Work & Health Conference*, Stockholm.

5A Industrial cleaning: safer at a lower cost

Contents

5A.1 Summary

Cleaning is an industry that appears to be expanding. In the not-so-distant past companies had their own cleaners but, increasingly, companies see cleaning as a better and cheaper service if provided by specialist enterprises. There has not necessarily been a change in the number of people employed in cleaning nor in the socio-economic sector from which the cleaners are drawn, rather cleaners have started their own small contracting enterprises.

This case study, although arising from an investigation into whether or not outsourcing is a better option than in-house cleaners, investigates the injury reduction and cost benefit of using modern management techniques and asking the question 'is there a better and safer way to do this job?'. The answer was clearly yes, which led to reduced costs and reduced injury risks. The management techniques employed included improved work organisation, suitable capital equipment and training for the cleaners.

5A.2 Who is the cleaning industry?

Traditionally the workers in this industry are non-skilled, being employed with a minimum, or even absence, of training. Even companies that specialise in cleaning are usually formed from people who started off as cleaners rather than people skilled in management.

Frequently cleaners are precarious workers, perhaps older and female. As far as physical effort and work stress are concerned they are similar to many workers in the hospitality industry (see case study 5B regarding hotel room attendants).

In a nutshell, cleaners are most frequently employed as if they were working on a small scale, and even office and shop cleaners use domestic or domestic-style equipment. Exceptions are for streets, railway stations and other large outdoor areas where sit-on industrial machines are used.

5A.3 Capital investment

Capital investment typically is low. The move to outsourcing has put pressure on costs, but there has been little movement away from domestic-style cleaning. The cleaners are left to manage as best they may as if cleaning their own homes. Vacuum cleaners and floor polishers are about the highest form of equipment for floor cleaning, but even there the equipment supplied may not match the requirements. Brooms, hand dusters, mops and buckets are the standard equipment for indoor cleaning.

5A.4 Costing for contracts

Part of the argument against increasing capital investment is the perception within the cleaning industry about the way profits are made. One method for costing a contract (for outsourcing) is to assess how long the work would take, and thus the labour costs, then add a percentage for overheads including equipment, materials and profit. This perception of profit as a percentage of wages prohibits an increase in labour efficiency, as reducing wage costs reduces the apparent profit! Not a logical way to run a business, but one that is not uncommon.

Clearly, the logical way to arrange a contract price is to look at the net cost of cleaning (area, materials and frequency), the additional costs of overheads (supervision, insurance, taxes, capital equipment and depreciation, and so on) and then to add the required profit margin.

If the market economy theory is correct (see section 2.4), the numerous small cleaning enterprises should compete for contracts, keeping prices to a minimum by using the most efficient equipment and work systems. There should be a reduction in costs as the cleaners compete for the outsourced work.

5A.5 Quality control

With in-house cleaners there is often a lack of expertise to supervise cleaners, so there is a tendency to leave the cleaners to their own devices and not to interfere with their work; they are 'invisible'.

With the rise of outsourcing one would expect that the client would be more aware of costs and may well demand more for their money, including quality control by the cleaning enterprise, but this is not always the case. Outsource cleaners may also be 'invisible' to both the client and the cleaning company as they frequently work outside 'office hours'. This leads to the possibility that quality may be sacrificed for price.

Working while their children sleep

Schools are a typical example where outsourcing may sacrifice the cleaners' health. Cleaners have to work outside school hours and start at 05.30 to finish at 08.30 and then start again at 15.30 to finish the day's work at 18.30.

In this example the school cleaners were mostly immigrant women with little command of the common language, English, and were under employment pressure (unemployment among this group of people was several times the national average). They had no means of redress, little contact with their employer, little or no knowledge of their employment rights and were 'free' of any likelihood of inspection by the labour inspectors. With high work pressure, poor or unsuitable equipment and generally under-resourced, it is little wonder that they suffered from musculoskeletal injuries to the back and upper limbs.

However, even with the financial pressures of the market economy there was no need for the cleaners to have suffered injury. The conclusion from this particular study was that better equipment, replaced as necessary, and more effective work organisation would have reduced the stress on the cleaners and reduced the injury rate. These conclusions included better determination of staffing levels, and better communication between the cleaners, employers and clients.

None of these conclusions is surprising but they are frequently found as factors in injury causation. In a nutshell, the injuries were due to poor management practices on the part of both client and employer and, in the long run, no money was saved by the inefficient employment of cleaners. The cheapest contract is not necessarily the one that is the cheapest in the long run (personal communication; C. Aickin).

This sacrifice is not just for quality, the cleanliness of the place cleaned, but for the workers themselves, their very health and safety.

5A.6 The workplace

The following case study is a typical example of management using cost-benefit analysis to increase productivity and reduce injury potential.

There was senior management pressure on the maintenance section of an organisation, a university, to change from employed to contract cleaning staff (from in-house to outsourced cleaners) as part of the prevailing fashion in the direction of outsourcing. The maintenance manager asked the question 'is it really cheaper to use outside contractors rather than our own cleaning staff? Our major criterion is that we need to maintain the same high quality of cleaning.'

As well as cost, to be discussed later, the maintenance manager used the argument that, as the primary contractor, the university would still have the full legal responsibility for the health and safety of the cleaners; whether the cleaners were considered to be employees or contractors mattered little as there would be no change in the legal, or for that matter moral, position.

The university was a self-insurer for workers' compensation and it was considered unlikely that there would be savings on workers' compensation costs: workers' compensation premium and injury costs would be reflected in any contract price.

Frequently a company would put out a contract and let those competing for the contract determine the lowest price; a basic tenet of the market economy. This is reasonable if only price is considered but if other aspects, i.e. quality and safety, are important then other criteria need to be used as well. In this case the maintenance manager determined for himself what a reasonable cost would be, taking into account the quality and frequency of cleaning and the safety of the cleaners. He was then in a position to answer the question 'is it cheaper to do the cleaning with our own staff or to outsource?'.

As a university there was a multitude of buildings but, for the sake of illustrating the principle, we shall only look at one building although the same principle was used throughout the university.

The building is an office/classroom/laboratory building of about 8,600 square metres (93,000 square feet). The basic methodology employed by the maintenance manager was to measure the area of each space (offices, corridors, toilets, etc.) along with the type of floor coverings and fittings and determine the standard and frequency for cleaning each area. There are standard tables within the industry that give time bases for cleaning offices, toilets and so on and then add an allowance for rest breaks.

From this the maintenance manager was able to determine a fair price for the contract and compare the quotations received. He was in a position to ask 'what equipment and chemicals does each contractor propose to use and how efficient will each be?' and not simply 'what is the price?'. For such a highly utilised area as a university building, public health is a major consideration and cannot be sacrificed for a low price.

The maintenance manager was putting in place his own knowledge of the results required and not simply relying on the contractors to do a reasonable job. He was, in effect, ensuring good management practice.

By this stage of the process the maintenance manager was in a good position to look at his own cleaning staff. There were, by now, several reasons to improve the situation, and not only to justify keeping the in-house cleaning staff *vis-à-vis* outsourcing: the injury rates of the present cleaning staff were unacceptable, being above the University average. To go to contract cleaners may increase the injury rate and, eventually, be an added cost to the university, with the net result that the university pays for the cost of injuries but has no effective control over their prevention. The same applies to the legal and ethical situation; responsibility but no control.

This is a bind that many companies find themselves in – paying for an injury cost and having a legal responsibility over which they have no control.

The management strategy for improving productivity and reducing injuries was twofold. It consisted of looking at the:

- work organisation of the cleaners (can their time be used more effectively?)
- individual cleaning tasks (can each task be done more efficiently and more safely?).

Costing cleaning contracts for safety and efficiency

Better management tools are needed to assist small cleaning companies; to ensure that they can manage a contract, not injure their cleaners and still make a profit. For many years there has been published material designed to assist in determining cleaning schedules, but this has mostly been used by larger companies. More recently interactive software systems have been introduced for determining reasonable work schedules: schedules which are reasonable to ask people to do safely and efficiently.

One such computer system has been developed by a cleaning materials and equipment supply company that also conducts training sessions for supervisors and cleaners. This computer system (for example, see www.interclean.com.au) allows details to be entered of the work areas, types of cleaning equipment, average time to clean such areas and the frequency of cleaning (see Figure 5.1). By adding the individual wages of the cleaners the gross annual cost for cleaning the area is computed. Thus this software becomes an aid to small contract cleaners for costing, quoting and ensuring safety and profit.

Figure 5.1 Having to move furniture and clean around furniture takes considerable time and effort, and extra time should be allowed for these additional activities.

5A.7 Work organisation

Over a period of some years the workloads of the cleaners for this building had become skewed; on some days the workload was light and on other days heavy. To even out the workload made better and more effective use of the time and, as work overload is a well-known risk factor for musculoskeletal injury, to even out the workload reduced the injury risk.

This is, of course, a relatively low-cost organisational improvement and is accommodated within the normal management cost structure. However, it still takes time to determine new methods and this is time away from other management duties; the cost of management time is included in the case study.

5A.8 Individual cleaning tasks

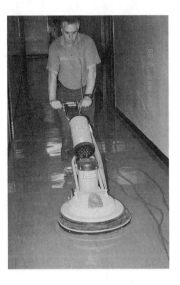

Figure 5.2 The design of the 'old' floor polisher gave a turning motion requiring force to maintain a straight path.

Figure 5.3 The improved polisher keeps to a straight line, with little force needed from the cleaner.

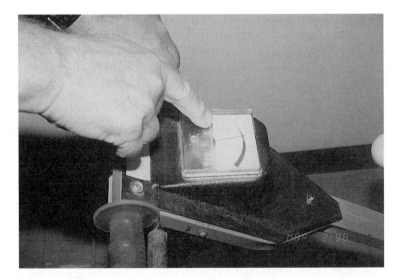

Figure 5.4 The improved polisher has a detection system so that the cleaner can maintain the optimum pressure.

Figure 5.5 A vacuum cleaner with a small head can clean only a small area at a time, and requires considerable pressure to ensure that it presses into the carpet.

Figure 5.6 A vacuum cleaner with a wide head can clean large areas quickly, and does not require the cleaner to use force or adopt stressful postures.

In this building the corridor and laboratory floors were all vinyl tiles requiring frequent polishing. The cheap and popular rotary polisher (Figure 5.2) was slow to use and, as there was a build up of polish, the polish had to be removed (stripped) at regular intervals. Stripping is labour intensive and stressful on the musculature.

The rotary polisher was replaced by a straight line polisher (Fiugre 5.3) which, in the model chosen, is three times faster per unit area than the rotary polisher. Moreover, there were unexpected quality gains; the floor did not need stripping each three months as it did formerly and, at the time of inspection, had not been stripped for three years – with no decrease in floor appearance.

As in all industries there are fads and fashions; in the cleaning industry small back-pack and pull-along vacuum cleaners (Figure 5.5) have become popular. These types of vacuum cleaners are very useful for small areas, particularly those with poor access such as staircases, under desks and tiered seating in lecture theatres, but their use had been extended to all areas even those with large expanses of carpet.

In the larger carpeted areas of the building these vacuum cleaners were replaced by floor model self-propelled vacuum cleaners (Figure 5.6) which are more efficient. The reason for the increase in efficiency is that the back-pack and pull-along vacuum cleaners have a small head and have to be pulled back and forth over the same area of the carpet to be cleaned; it does a 'spot clean' rather than a full clean; the floor model has a wider head and, being self-propelled, the worker just follows it over the area for one sweep. Simply by going from a 9″ (23 cm) width head of the back-pack cleaner to 27″ (69 cm) width head of the floor model provided an increased cleaning efficiency and a saving in time of not less than three fold.

As a bonus, removing the back-pack from the backs of the cleaners removed a potential source of injury.

5A.9 Working and exposure time

Previously it took 27 man-hours each day to clean this particular building. After the work reorganisation and equipment changes, the time was reduced to 19 man-hours each day; effectively a reduction from four to three cleaners per day.

Although there were clear productivity and cost gains due to the reduction in time required to clean the building, was there a concomitant reduction in injury risk? After all, three people were now doing the work of four.

Musculoskeletal stress and injury has several causes, including load, daily duration of exposure to the load, posture of the individual and so on. Reduction of the time spent at any particular task implies a reduction in exposure and thus a reduction in injury risk.

The time the cleaners now spend with a small head vacuum cleaner is reduced to only a few difficult-to-clean areas; use of the self-propelled vacuum cleaner imposes no musculoskeletal stress on the user. Similarly, the change to the straight line polisher reduced the time spent using the machine by two-thirds and reduced significantly the exposure to musculoskeletal stress. With the organisational changes eliminating overload periods by evening out the daily tasks, it is clear that the risk of musculoskeletal injury had been reduced considerably.

5A.10 Training

There is pressure on the cleaning industry to be more efficient. The use of higher quality and more expensive equipment implies that cleaners have to match their work techniques with the equipment. Cleaners are no longer 'domestics grown large' but people who have to know their jobs, for which they need training.

Training has to include reduction of hazards to the cleaners as well as increasing productivity and efficiency and, for the contract cleaner, profit. For example, the number and types of cleaning materials are increasing, with varying degrees of toxicity, and machinery is becoming more powerful and hence more potentially hazardous. Moreover, there is increasing pressure on the cleaners to work harder and faster.

The university case study above illustrates that it is possible to increase productivity and efficiency yet reduce physical stress and reduce the likelihood of a musculoskeletal injury. However, it is clear that such changes can only go so far and further change may lead to injury. The reduction from four to three cleaners each day does not mean that the number can be further reduced to two cleaners without a reduction in quality of work and an increase in injury rates.

Training must include two interrelated issues:

- the improved use of equipment with better cleaning techniques
- the means to avoid injury – working safely.

Table 5.1 Replacements of old cleaning methods by more efficient and safer methods

Old methods	New methods	Health effects	Productivity changes
Short handle toilet brush	Long handle toilet brush	Reduced bending	Faster use
Fixed length vacuum cleaner wand	Adjustable length	Reduced bending	Faster use
Wet mop	Micro-fibre mop head	Reduce dust	Faster; only one pass needed
Chemicals diluted by guesswork	Chemicals diluted by measurement	Less exposure to chemicals	Reduced chemical usage
Wastepaper bins under desk	Office occupiers put waste bins in corridor	Reduced bending	Not doing unnecessary tasks
No training	Training	Less stress, better health	Improved productivity

Thus, the old-fashioned cleaner, who started at the bottom and worked his or her way up to be a supervisor or to own his or her enterprise must also be trained not only in modern cleaning techniques but also in management. This is starting to be realised in the larger contract cleaning companies and the larger organisations that maintain in-house cleaning staff, and training by specialist trainers is more common than before.

An essential component of training is information about efficient equipment, materials and work techniques that is translated into work practices. An enterprise that one of the authors worked for many years ago had a training course for its cleaners given by an outside training company; however, it was made clear to the cleaners by their supervisors (after the trainers had left) that some of the cleaning materials demonstrated, although clearly of use, would not be supplied as they were 'too expensive'. It made a mockery of the training.

As well as purchase of the equipment mentioned in the case study, other changes had also occurred to improve efficiency and reduce injury risk. Table 5.1 gives some simple means by which efficiency was improved throughout the university at comparatively little cost. It included training the cleaning staff (measuring chemicals) as well as cooperation by other staff members (placing the wastepaper bins in the corridor).

5A.11 Cost-benefit analysis

Following is a summary of the analysis made by the maintenance manager of the situation before he made the changes (the 'initial case') and the situation after the changes ('improved equipment'). The full analysis with notes is given in the CD of the productAbilityBasic program accompanying this book (the file will be found in *File\Examples\Cleaning.bgl*).

The cost of the intervention was $10,800, which included management time and new equipment. The reduction in the number of cleaners from four to three each day allowed some time spare each day for preparation. The costs per productive hour increased slightly due to the extra training required, but this illustrates the point that hourly costs can be misleading as the total employment costs were reduced from

Table 5.2 Cost-benefit analysis for the improved equipment and work system

		Initial case	Improved equipment
Number of full-time cleaners		4	3
Paid time per cleaner (38 hour week)	hour/year	1,976	1,976
Time paid but not worked (vacation, illness, etc.)	hour/year	364	372
Productive time (actual time worked)	hour/year	1,612	1,604
Wage paid directly to each cleaner	$/hour	14.13	14.13
Employment cost (wages plus overheads of supervision and administration)	$/hour	17.81	17.81
Productive employment cost (employment cost per productive hour)	$/hour	21.83	21.94
Fixed employment cost for the cleaners	$/year	140,756	105,567
Reduced productivity (29%)	$/year	40,819	7,654
Total employment cost for the cleaners	$/year	181,575	113,221
Intervention costs (cost of equipment)	$/year	–	10,800
Savings (assumed to occur in one year)	$/year	–	68,000
Pay-back period	months	–	2

The University was self-insured and workers' compensation insurance costs have not been included. See productAbilityBasic; File\Examples\Cleaning.bgl.

$181,000 to $113,000 per year (a reduction of 37%). The pay-back period was two months, with savings of $68,000 per year (see Table 5.2).

An argument can be made in this case that the reduction in hours required to clean the building is the increase in productivity and that the item 'Reduced productivity' is double counting. It depends on the factors that are acceptable to management and included in the various cost items. The reduced, or lost, productivity due to unsuitable equipment (in the initial case) was identified as a factor contributing to increased costs; the reduction in hours to clean the building each day was a consequence of this finding.

However, a sensitivity analysis in this particular case study shows that if we ignore completely 'Reduced productivity' in both the initial case and the improved equipment case (reduce the tables to no loss; you can do this in productAbilityBasic; see *File/Examples/Cleaning.bgl*), the savings are reduced by about one-half to $35,000 per year and the pay-back period is four months. This is still an attractive result and indicates the way in which the cost-benefit model can be used to illustrate cost-sensitive parameters.

5B Permanent or precarious employment? The hospitality industry

Contents

5B.1 Summary

This case study illustrates a service industry where the cost-benefit analysis showed a clear benefit when an ergonomics intervention was introduced to relieve musculoskeletal injuries.

In the service industries there is a tendency at present to employ people on a 'precarious' basis, which has led to reduction in the quality of working conditions and a rise in injury rates (Quinlan *et al.*, 2001). Many companies believe that using contract workers or self-employed workers will relieve them of their responsibility for the care of their workers. Such a view cannot be substantiated ethically, legally or in cost terms.

5B.2 Precarious employment

The terms 'precarious', contingent' or 'just-in-time' workers have been coined as a category to include part-time, casual, temporary, contract workers and self-employed workers. Until recently it has been thought that the 'normal' type of employment was full-time work but, for a large proportion of the workforce, this has rarely been the case. Full-time employment was normal for the male working population in developed countries, with exceptions in the building and some other industries, but full-time employment was not normal for a large proportion of the female working population. In particular, the service industries have attracted part-time employment and, most often, female workers. This tendency to part-time employment seems to be increasing.

The industries where precarious employees are found include the fast-food and supermarket industries, warehousing, cleaning and hospitality services, the last of these including restaurants, hotels and tourism. The requirement for domestic and health care workers is increasing with the ageing of the population, and these workers are mostly 'precarious'.

For the hospitality industry, needless to say, good service is the hallmark by which a company survives or goes under. Nowhere is this more apparent than in the hotel industry where the more prestigious the hotel (the greater the number of 'stars'), the greater is the customers' expectation for good service.

5B.3 Occupational health and safety responsibilities

This next case involves a four-star hotel that had seen its workers' compensation premium rise to such an extent that was affecting the overall company profit. High standards had to be met but so did profit, and it seemed that the two were becoming incompatible.

Firstly, the reasons for the rise in workers' compensation had to be identified. This was the easiest part of the exercise – musculoskeletal injuries to the back and upper arms of the room attendants were the major part of the costs for the injuries.

The room attendants, or chambermaids as they used to be called, were almost entirely female, either older women, and thus not as strong as they used to be, or from South East Asia – mostly small and slight. For heavy work one would preferably pick young, strong men but, as it is neither practicable, ethical nor legal to choose in that fashion, the hotel was left with a physically weaker proportion of the working population. This is by no means an unusual situation in industries that include manual work but have a large proportion of female workers, e.g. nursing and health care, cleaning and service industries.

There were several options open to the hotel to reduce its workers' compensation costs. One method was to employ more staff from an outside agency rather than its own staff. Of 96 room attendants, 66 were full-time employees and 30 were contract staff from an outside agency. With a turnover of 60% per annum of the full-time staff, it would be quite simple to substitute full-time staff with contract or precarious labour.

There is a mistaken belief that if workers are self-employed or employed by an outside agency the company is not responsible for their health and safety. This legal aspect may need testing in some jurisdictions but, in this particular jurisdiction, the responsibility for the health and well-being of workers is the responsibility of both the employer and the occupier of the premises: with precarious workers the employer is the employment agency but the occupier of the premises is the hotel. The hotel still has the responsibility for all those on its premises, including workers whether they are its own employees or contractors. In other words, responsibility for health and safety cannot be shed by employing contract labour rather than its own employees. Shedding responsibility cannot be substantiated ethically, legally or in cost terms.

Even without the legal obligation to ensure a safe and healthy workplace, the cost to hire contract staff is greater than the cost for its own employees – somebody has to pay for the workers' compensation, running costs and profit for the agency. The hotel did try to increase its ratio of contract workers to full-time employees but the strategy was not successful; as well as increased costs there were difficulties with training and quality maintenance. It was not easy to maintain quality of service with changing contract staff who did not have a long-term commitment to the hotel.

5B.4 The workplace

The main factors that led to musculoskeletal injuries in the room attendants were found to be:

- stretching when cleaning the bathroom tiles, mirrors and windows
- force and repetitive actions in vacuum cleaning
- bending and force when moving beds.

The photographs in Figures 5.7–5.11 illustrate these factors.

Figure 5.7 Moving a bed with the knees is safer than bending and pushing or pulling it away from the wall.

Figure 5.8 Correct tools would eliminate the need for a person of small stature to stretch and reach far above her head.

Other factors that were identified as leading to injuries were the organisational and time pressures. The organisational pressures included loading the trolleys at the pantries, which often were not even on the same floor as the rooms to be cleaned, and the short times allowed to clean each room. Workloads were uneven; some days had high workloads (many rooms to be prepared for new hotel guests) and some days had low workloads.

Although outside ergonomics specialists were brought in to advise management on injury reduction and hence cost reduction, the method chosen by the ergonomists was to let the room attendants determine their own better and safer methods. Working committees were set up consisting of representative room attendants and the hotel's occupational health and safety manager.

After the committee sessions the members of the committee were able to identify the work tasks that lead to a bad back but had considerable difficulty in identifying

Figure 5.9 Installation of furniture and fittings not designed for easy cleaning makes the task hazardous as well as inefficient.

Figure 5.10 Cleaning a bath requires much bending.

Figure 5.11 Cleaning in the bath is one way to reduce bending but is possible only for a person of small stature.

tasks that led to more subtle forms of musculoskeletal injury, particularly to the upper limbs. The occupational health and safety manager believed that this was because a backbone model that visually illustrates back injuries was used but there are no models that adequately illustrate chronic upper limb injuries.

There is also general knowledge about a 'bad back' and, in any case, the person will usually feel the onset of a back strain but not usually the onset of a shoulder strain. It was in the latter area that the occupational health and safety manager had to give extra support and knowledge to the committee.

The committee identified workplace training as an aspect that required considerable improvement, and the training manuals were rewritten to include the methods identified as safe. The occupational health and safety manager ran the training sessions for up to six room attendants at a time. The training rooms used were actual hotel rooms to allow for a hands-on training mode.

Other items identified by the committee and implemented by the hotel management included:

- better trolleys and vacuum cleaner heads
- trolleys and vacuum cleaners regularly maintained
- new employees correctly trained
- collection of the used linen from the trolleys by an extra employee, which lightens the trolleys and reduces the need to wheel the trolleys to the pantries.

Organisational changes reduced injuries among part-time workers

Part-time (precarious) workers were suffering severe musculoskeletal injuries to the neck, back and shoulder from working in an area manufacturing high-voltage equipment. There were four work stations, each employing one person working for six hours each day. Job rotation was not practised and each employee worked only at a single work station.

Analysis of the jobs determined that the musculoskeletal injuries were caused by the static and/or dynamic loads imposed on the workers while they were performing some of the tasks. Although the work stations had been upgraded by standard physical ergonomics methods, the improvements had had no effect on the injury rate.

The assembly line productivity was low and, as new investment could not be justified in economic terms, an alternative non-capital investment was tried – a change in work organisation. A fast schedule job rotation was introduced and employees changed work stations and tasks once per hour. The idea behind this arrangement was that, although each job gave stress to certain muscles, the muscle groups under stress were *different* in each task. The set of muscles used at one work station was 'rested' at the subsequent work station while a different set of muscles was active. Each change would provide a recovery period for the muscle groups under stress from the job at the previous work station.

The incidence of the part-time employees' musculoskeletal injuries after the job rotation scheme was introduced dropped to zero and there was no loss in productivity.

As well as the elimination of work-related injuries, it was noted that there was a reduction of 50% in the non-work-related sickness absence rate. The reason for this was not investigated.

One important change to the work methods was in moving beds. As with most tasks there are easy and difficult ways, and the new method was to ensure that the room attendants used their knees to push the beds rather than bending and using their backs to pull the beds.

There can be little doubt that by the room attendants pooling their ideas and the occupational health and safety manager introducing their best and safest methods into the training, the training sessions were a most effective form of injury reduction.

The hotel decided not to replace its room attendants with contractor labour but to reduce the number of full-time staff (eight-hour shifts) and increase the number of casual staff (five-hour shifts). Contractors are only used at times of overload. However, during the duration of the investigation (two years) the number of full-time staff was only reduced from 66 to 55 and thus the increase in part-time staff would not, of itself, account for the large reduction in injury rates and costs.

The change to casual staff has affected where the room attendants come from (their socio-economic group). There is a move towards 'mothers-at-home' who want some part-time work to supplement the family income. Whether or not part-time work led to some of the room attendants taking other jobs and thus increasing their actual risk of injury is a question that we did not investigate and, indeed, was not investigated by the hotel.

5B.5 Cost-benefit analysis

Table 5.3 shows the data for the case study before and after the changes (the 'Initial case' and the 'New conditions' respectively). The overall staffing, calculated as the number of full-time equivalents, remained the same in the two situations.

These work procedures and organisational changes led to a marked decrease in both the incidence of injuries and the injury severity. The reduction in injury absence led to an increase in productive hours worked, with a decrease in productive hourly costs for the full-time room attendants.

The extra training received by the room attendants enabled them to achieve a higher quality of work which, in turn, required less supervision, leading to a reduction in the number of supervisors. There was also a reduction in labour turnover from 60% to 40% per year; these factors are noted in Table 5.3 as 'Other costs'.

The workers' compensation insurance costs came down substantially. Although the industry tariff rate remained unchanged (set by the government actuaries as a percentage on wages), the injury experience payment came down from $138,000 to $20,000 in the first year.

The intervention cost of $96,000 was paid for by the reduction of $118,000 compensation insurance costs in the first year. The total annual saving was $815,000 which indicates a pay-back period of only 6 weeks.

Seemingly there is a cost saving when employing casual instead of full-time employees but, as noted in Chapter 3, one has to be careful not to assume too much accuracy as the figures depend on the information entered and how complete the data is. The 39 cents per productive hour difference between the full-time and casual room attendants is less than 3% of the productive cost and may not be significant.

This is an excellent example of a workplace situation whereby the aim was to achieve a reduction in injuries and injury costs. Not only was that aim achieved but, by management and workers assessing the work together, it enabled other benefits to

Table 5.3 Injury reduction of room attendants and cost analysis for employing full-time and precarious workers

	Initial case		New conditions			Comments
	Full-time	Contractor	Full-time	Contractor	Casual	
Numbers of room attendants	66	30	55	8	73	
Full time equivalents (FTE)	66	30	55	8	33	
Paid hours/year	1,976	1,976	1,976	1,976	900	
Total paid absences hours/year	459	–	376	–	–	
Productive hours/year	1,517	1,976	1,600	1,976	900	Takes into account vacation and sickness absence for full-time and paid for by the hotel. Contract and casual workers do not get paid vacation and sickness absence. There is a 20% loading for casual employees as they get no paid illness or vacation leave
$ Wages/hour	11.00	28.00	11.00	28.00	13.20	
Productive employment cost ($/hour)	14.33	28.00	13.59	28.00	13.20	Hours worked and paid for by the hotel
Total employment cost ($/year)	1,434,000	1,660,000	1,195,000	442,000	867,000	
Other costs ($/year)	672,000		468,000			Includes supervision, administration and turnover (recruitment and training)
Workers' compensation	34,500		35,100			Tariff rate
Cost ($/year)	138,000		20,000			Injury experience rate
Intervention costs ($)	–		96,000			
Total cost ($/year)	3,938,000		3,123,000			
Savings ($/year)	–		815,000			
Pay-back (weeks)	–		6			

be achieved; to the yearly cost savings was added an increase in quality of work, an essential part of running a high-class hotel.

5B.6 Reference

Quinlan, M., Mayhew, C. and Bohle, P., 2001. The global expansion of precarious employment, work disorganisation and occupational health: a review of recent research. *International Journal of Health Sciences*, 31, 335–414.

5C Why lift patients when there's a better way? Reducing back injuries in the health industry

5C.1 Summary

An editorial in *The Lancet* (1965) sums up the dilemma of manual handling in the health industry, be it in acute or residential care:

> The adult human form is an awkward burden to lift or carry. Weighing up to 100 kg or more, it has no handles, it is not rigid, and it is liable to severe damage if mishandled or dropped. In bed a patient is placed inconveniently for lifting, and the placing of such a load in such a situation would be tolerated by few industrial workers.

Because of the awkwardness of lifting, patient handling has long been identified as the most common cause of back injuries for nursing staff (Stubbs *et al.*, 1983). Appropriate methods to reduce the incidence of injuries are well known, and Swedish researchers identified that lifting work was reduced by approximately 50% through well-designed wards with sufficient space to accommodate lifting equipment (Ljungberg *et al.*, 1989). This knowledge of prevention methods has not been sufficiently utilised, nursing staff are still being rostered to work alone when lifting patients, and hospitals and nursing homes are still being built and refurbished without considering sufficient lifting equipment.

Often the initial cost of these improvements is cited as the reason that they are not undertaken. However, an increasing number of hospitals and nursing homes are tackling these problems and showing that the changes can be cost-effective. Fortunately, in some countries hospital design to take into account proper patient handling facilities is obligatory.

This case study concerns a typical nursing home which is a privately funded, non-profit making organisation working on a fixed and restricted budget. As its workers' compensation costs increased, so the concern of management increased with the realisation that the costs needed to be contained and reduced. Investigation soon identified the major source of the compensation costs – back injuries among the nursing and caring staff.

The nursing home tackled the problem systematically by undertaking a risk assessment which included asking the nursing and caring staff for their views of the hazards. Patient lifting was assessed as the main hazard, with lifting patients in the toilets identified as the largest single problem. Many of the nursing home's 70 elderly patients were not able to walk and, due to their relative immobility, they had to be lifted from the bed to the chair to the toilet to the shower, and so on.

Traditional means of back injury prevention include installing mechanical lifting equipment, ensuring sufficient space in which to use the equipment, and training staff in its use. Introducing these measures led to improved efficiency, reduced injuries and reduced workers' compensation premiums. The increased efficiency allowed restructuring of the staffing rosters to ensure that no nurse worked alone; this further reduced injuries.

5C.2 The first essential – the right lifting equipment

A risk assessment was undertaken and the results were in accordance with ergonomics research; that lifting patients was the major hazard. The director of nursing determined that funds would need to be spent on both equipment and patient areas (bedroom, bathrooms) to minimise manual lifting of patients (see Figures 5.12–5.16):

- sufficient mobile patient lifting equipment was purchased, together with a sling for each patient who needed the equipment
- manually adjustable beds were replaced by electrically adjustable beds
- bathrooms and toilets were enlarged
- existing equipment was modified for ease of movement.

To ensure they purchased equipment that best suited their needs, prior to purchasing the nursing staff tried out different mobile patient lifters. During this trial period the reasons for nurses not using lifting equipment were identified, e.g. when the lifter or an appropriate sling was not readily available. This helped determine the number of each type of patient lifter required and that each patient required their own sling.

Another problem identified in the risk audit was that nurses worked at inappropriate heights through not adjusting the manually adjustable beds. It was concluded that, as electrically height-adjustable beds are more easily adjusted, nurses would be more likely to adjust them and so work at a suitable height.

Some of the equipment modifications made were innovative and came from 'looking at an old problem through new eyes'. When disabled patients got up in the morning they had to be lifted from the bed into a wheelchair, from the wheelchair to a shower chair, from the shower chair back to the wheelchair, from wheelchair to dining table chair, from the dining table chair to the wheelchair, and finally from wheelchair to lounge chair. When put like this it becomes obvious that a single mobile chair could be used and, in fact, the lounge chairs were put on wheels. Except for

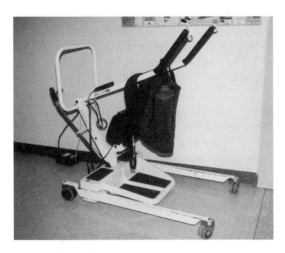

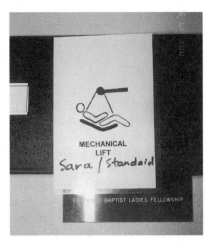

Figure 5.12 Providing sufficient patient lifters and training reduces the amount of manual effort required by the nursing and caring staff.

Figure 5.13 The physiotherapist determines the most suitable means for moving disabled patients, and places a visible symbol above each bed for the carers to follow.

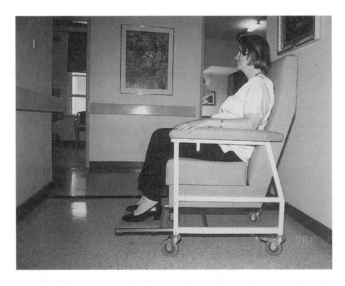

Figure 5.14 A comfortable armchair is a simple means to reduce the number of transfers required for each patient.

hygiene (showering and toilet) requirements, the patients are able to be in the lounge chairs for most of the day and the chairs were even able to be used in the garden for outdoor activities.

For a similar reason the commodes were put on wheels so that the staff could wheel the commodes rather than carrying them from room to room.

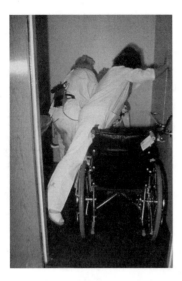

Figure 5.15 Manual handling in a cramped space, in this case a toilet, makes back injuries almost inevitable for the carers and is not dignified for the patient (from Ljungberg *et al.*, 1989).

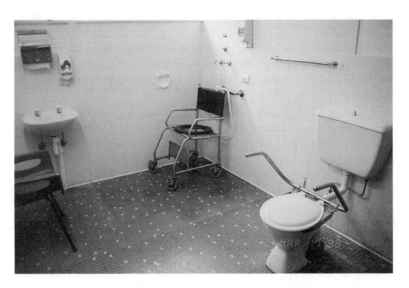

Figure 5.16 Enlarging the area by removing the dividing wall between the toilet and bathroom enables the carers to work safely.

From discussion with the nurses it turned out that, due to the small size of the toilet rooms, the toilet arrangements were almost 'designed' to ensure the utmost difficulty and back injury risk for the staff. It proved possible to enlarge the toilets by knocking out the walls between the showers and the adjoining toilets.

With the larger toilets the staff can work with considerably reduced risk to their backs, with the additional important benefit of allowing more dignity for the patients.

Table 5.4 Equipment costs in the first year

	Description	Cost ($)
New equipment	40 slings for lifters ($250 each)	10,000
	3 extra lifters ($2,500 each)	7,500
	20 electric beds ($1,600 each)	32,000
Modify existing equipment	20 lounge chairs on wheels ($500 each)	10,000
	16 commodes on wheels ($50 each)	800
	toilet/shower conversions	4,050
Cost of the changes		64,350

The female nursing staff wore A-line dresses, which restricted movement and prevented them having a good lifting posture. A new uniform was introduced which incorporates a divided skirt or trousers which freed up movement, allowing the nurses to adopt a correct lifting posture.

The costs for the equipment and modifications are shown in Table 5.4.

5C.3 Getting the staffing right

Given the capital investment that had been made in equipment, it was critical that staff use the equipment and use it correctly. Training was a major factor, and the nursing home's part-time physiotherapist trained all nursing staff in lifting, including the use of lifting equipment; newly employed nurses or carers, on any shift and at any level of skill or experience, have to be trained by the physiotherapist before they are allowed to lift, or assist to lift, a patient.

To assist in reducing back injury risks the physiotherapist assessed the lifting requirements for each patient and attached a sign to each bed indicating the lifting requirements for that patient. All supervisory staff are required to ensure that the nursing and caring staff follow these signs.

Previously all patients were showered in the morning before breakfast (in a two-hour period), which meant a rush on the limited facilities and undue haste for the staff. Haste increases the stress on the muscles and increases the risk of a lifting injury; lengthening the period for showering patients to all day, with some patients being showered in the morning and some later in the day, reduced the injury potential.

These measures enabled better rostering of staff over all shifts, ensuring that there are always two nurses available for any lift, even at night; staff now work in pairs rather than singly.

With the reduced time spent on patient transfers and the time spent 'finding' another nurse, substantial gains have been made in the quality of service and care:

- each patient is now seen by the nursing staff at least once in each two hours
- more time is spent on social activities such as taking the patients into the garden.

The nursing home is now close to having eliminated manually lifting patients, but are there cost savings for the nursing home?

Table 5.5 Benefits to the nursing home in the second year

Details	Old system ($)	After the modifications and new equipment ($)	Difference ($)
Workers compensation premium	324,000	168,000	156,000
Intervention cost			64,350
Savings due to the changes			156,000
Pay back period (months)			5

1. Management and training time has not been included but it would be relatively small compared with the equipment and modification costs. The cost of the new uniforms is not included as a cost to the nursing home, as this was borne by the staff from their existing clothing allowance.
2. The Productivity Assessment Tool assumes that all costs and benefits occur in the same year. In this case study the reduction in workers' compensation premium did not occur until the expenditure program was under way and there was a reduction in injury costs.

5C.4 Costs versus benefits

Nursing homes are no different to any other industry – to save back injuries, minimise manual lifting! By following this philosophy the nursing home has achieved good results in a very short time.

Table 5.4 shows the equipment bought during the first year. It should be noted that this was all done within the fixed nursing home budget; monies were not conjured out of a hat.

By changes in the staffing arrangements the lifting frequency (lifts per hour) and the lifting exposure have been substantially reduced. In fact, all lifts are two-person and/or use a lifting machine so that lifting stress has been reduced and is on the way to being eliminated.

Nursing staff notice that they are not exhausted when they go home; this has led to an increase in morale and nursing staff often say 'they don't know how they managed before'.

Table 5.5 shows the cost-benefit analysis of the program to reduce manual lifting; the reduction in back injuries led to a reduction in the workers' compensation premium from $324,000 to $168,000 per year. This reduction of $156,000 in the workers' compensation premium shows a pay-back period of five months. As the expenditure was all within budget, the savings have been considerable and continue year by year.

5C.5 Design and redesign

When a nurse or carer is manoeuvring a patient onto or off a toilet in a cramped space, or holding open a door while trying to push a wheelchair through, or manually lifting when there is no extra person to assist, there is an injury risk. In this case study the introduction of better equipment and design of patient areas was an addition long after the building had been constructed. It was more by luck than by planning that mobile lifting equipment could be accommodated in the existing bedrooms and that the toilets and showers were able to be modified into larger rooms.

It is usually assumed that to fix a problem is more expensive than to prevent it in the first place but, in the case of building or rebuilding hospitals and nursing homes, providing the extra space needed to accommodate safe handling of patients appears to be more expensive. Additional space is required around each bed for lifting equipment, bathrooms need to be larger, and shared spaces (lounges and dining rooms) need to be considered. Doorways and corridors need to be wide enough to allow patients to be moved in their beds and space provided to store equipment (WorkCover Safety, 1999).

Although each additional square metre of floor space adds to the initial building costs, the substantial savings in injury costs should be considered against the cost of providing this extra space. As there will be several years between allocating finance for construction of a new hospital or nursing home and the date when it is occupied, the benefits gained will lag behind the costs. Comparing the additional building costs with projected reductions in injury costs can provide an estimate of the cost-benefit and Chapter 2, in particular section 2.6.2, indicates methods required for long-term assessment of the costs and benefits.

Nursing homes with single bedrooms, attached bathrooms and overhead patient lifting equipment are about 25% more expensive for the initial building costs than the older homes of two or four beds to a room, fewer bathrooms and no overhead patient lifting equipment.

Choosing an appropriate factor for the estimation of a reduction in injury is critical. Research by Ljungberg *et al.* (1989) indicates that 50% of back injuries for nurses relate to patient lifting, so it is reasonable to assume that good design and equipment will reduce back injuries by 50%. Our experience shows that, in most cases, paying attention to good ergonomics will result in greater reductions than this.

In the case of this nursing home, its workers' compensation insurer estimated that, for a serious back injury, its costs would be $300,000; this sum covers medical and surgical costs and a pension over the life of the injured person. For a 60-bed nursing home the cost differential between a well designed and equipped building and the older style building is approximately $450,000 so that, over the life of the nursing home, only two serious back injuries need be avoided for the new design to be financially beneficial.

In this book we have looked at interventions to existing workplaces; to use the costs of injuries *avoided* in a new building is not well based in logic (one cannot foretell the future) but it can be a telling argument when used alongside actual injury experience and research into back injuries for nursing staff. Put the other way round, as we are sure from experience that reducing the degree and frequency of lifting prevents back injuries in existing buildings, to *design* a building to reduce patient lifting should similarly reduce back injuries and associated costs.

5C.6 References

Ljungberg, A.S., Kilbom, Å., and Hägg, G.M., 1989. Occupational lifting by nurses' aides and warehouse workers, *Ergonomics*, **32**, 59–78.

The Lancet, 1965. The nurse's load (editorial), 28 August, 422–423.

Stubbs, D.A., Buckle, P.W., Hudson, M.P., Rivers, P.M., and Worringham, C.J., 1983. Back pain in the nursing profession. Part I. Epidemiology and pilot methodology, *Ergonomics*, **26**, 755–765.

WorkCover Safety, 1999. *Designing Workplaces for Safer Handling of Patients/Residents* (Melbourne: WorkCover Victoria).

5D Large-scale experiments: valuable but not easy to carry out

Contents

5D.1 Summary

In this chapter we will consider two large-scale experiments that, although proposed for different reasons, had similar experimental designs. In the first example retail stores had wanted to reduce their injury costs through the introduction of a physiotherapy service and compared two groups of stores within the same enterprise; in the second example an employers' engineering organisation wished to see if there were benefits to be gained from a multi-skilling style of manufacture compared with a Taylorist style and made the comparison between two enterprises.

Both experiments were valuable in themselves but suffered from the serious difficulty of controlling for confounding factors. As discussed in Chapter 2, as the situation becomes larger and more complex, more assumptions have to be made and more confounding factors will encroach on the clarity of the results. These types of workplace experiments (or investigations) are essential but not easy to perform.

5D.2 Are workplace health and social services cost-effective?

There is considerable contention as to whether or not health and social services provided at work will improve the health of employees or whether this is an industrial relations exercise; moreover, is it a cost-effective activity for the employer? Such health services include counselling, a gymnasium, on-site physicians and nurses and other health practitioners, and other means to increase physical activity and well-being. The following study describes how a retail chain store determined whether or not the introduction of a physiotherapy service would be effective in reducing both injury absence and overall costs. Similar costs and benefits could be collected to decide the effectiveness of other health or social services, although we would note that there are other values to be taken into account as well as financial ones.

The rationale behind choosing physiotherapy as a service to its employees in this case study, rather than another service, was with regard to the manual handling injuries suffered by the employees. As considerable efforts had already been made to reduce and control manual handling injuries the injury rate was not high but it was still an unacceptable cost in a highly competitive industry. An added stimulus to reduce the injury rate was that the enterprise was self-insured; the initiative for the physiotherapy service came from the Risk Management and Insurance Department.

In conjunction with a physiotherapy/ergonomics consultancy, the enterprise decided on a pilot study. Three stores were chosen and the service provided for a period of one year.

In this pilot study there were no control stores for comparison of the results, so that the measure of the effectiveness of the study was dependent on the views of the store managers and the store staff. The results were encouraging enough to convince the enterprise that a full-scale experiment was worth trying but, to evaluate the effectiveness fully, control stores had to be included.

The experimental design for the full-scale experiment

The retail sector is highly competitive and cost data is very sensitive, so we have not used the confidential data from the enterprise. We have used the results derived from the data to illustrate the value of this cost-benefit approach, but without compromising either the outcome or the retail enterprise.

The 'ideal' experimental design would have been to compare matched stores based on similarity of store size, ratio of employment of full- to part-time staff and similarity of the socio-economic structure of the customer base. Due to the structure of the enterprise and the method of collecting statistics, this 'ideal' experimental method was not practicable and a 'compromise' experimental design was devised comparing 17 stores in one district (the trial or intervention group of stores) with the same number of stores (the control group of stores) in another district in the same city.

The intervention was the introduction of the physiotherapy service. A physiotherapist attended each store on one occasion each week for as many hours as required, usually half a day. During that time staff came to see them about any musculoskeletal problems they had, and received appropriate advice and treatment. The cause of the discomfort need not be work-related but if the physiotherapist thought that the cause was work-related, or that work was aggravating the condition, they would go to the work station and advise on changes to alleviate the cause (see Figures 5.17 and 5.18).

Comparing the control and trial group of stores for the same year for unplanned or illness absence due to musculoskeletal discomfort, such as low back pain, should measure the effectiveness or otherwise of the physiotherapy service. Unfortunately, it was not practicable to separate out musculoskeletal discomfort from other absence reasons, so that total unplanned absence was used as the basis for comparison.

We were able to collect data for two years, the year of the experiment and the previous year, for both the trial and control groups of stores. Thus it should be possible to:

- (*Experiment A*) compare the data from the trial group of stores over a two-year period, with and without the physiotherapy service (internal control)
- (*Experiment B*) compare the trial group of stores with the control group of stores for the same year (external control).

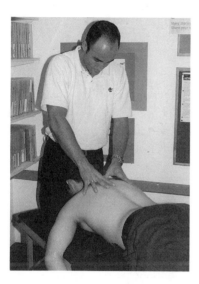

Figure 5.17 On-the-spot treatment of staff by a physiotherapist, with subsequent discussion, helps to identify potential causes (whether at work or at home) of musculoskeletal discomfort.

Figure 5.18 Assessment of risk, risk control and suitable training are essential parts of the physiotherapy service.

The advantage with Experiment A is that it overcomes some of the variation in the socio-economic structure of the customer base and other differences between stores. However, it has the disadvantage that it does not control for differences in the business cycle, general population illnesses and so on. These latter points should be controlled for in Experiment B.

The experimental period lasted for eight months in each of the two years of the experiment, and tracked about three million hours worked by employees in each group of stores in each year.

Confounding factors

In the year prior to the experiment the enterprise had changed the way that sickness absences were managed in the trial group of stores. Previously, an employee only had to telephone the store and tell a clerk that they would not be in on that day; the new system was that an employee had to speak directly to the section or store manager and state whether or not they planned to go to a doctor or how they planned to return to work as soon as possible. Returning to work quickly is cost-effective as this employer pays for illness absences, for part-time and full-time employees, and absences are covered by overtime by other employees; a double cost for absence.

The new system of controlling absence (the 'absence reporting system') was implemented in the trial stores the year prior to the experimental year and in the control stores at the start of the experimental year. This introduction of the new system of managing sickness absence at different periods in the trial and control groups of stores was a confounding factor, but recognising the confounding factor meant that we were able to take it into account when interpreting the results.

Another confounding factor was that at the commencement of the experimental period the trial group of stores had a significantly lower absence rate than the control group; this may have been due to the introduction of the sickness absence management system in the trial group in the previous year or some other, unknown, factor. Nevertheless, it is usually harder to reduce absences when the rate is already low than when the rate is high.

As we have noted in Chapter 2, the more complex the situation the more difficult it is to interpret the results. A cost-benefit analysis of a simple system, for example the introduction of the automatic darkening welding helmets described elsewhere in this chapter, is easier to interpret and understand than a complex system beset with confounding factors, many of which one may not be able to identify, let alone measure.

Results

The results discussed in this case study are only for unplanned (illness) absences and did not include measures of productivity. Sales per employee or sales per employee hour would be good measures of employee productivity, but the computer recording system did not allow these measures to be extracted in sufficient detail.

One needs to use data cautiously and illness absence rates would be expected to be a reasonable reflection of the effects of a physiotherapy health service. However, 'unplanned absence' data is for all absences whereas one would expect the physiotherapy service to have an effect specifically on musculoskeletal absences; thus the

Table 5.6 Experiment A. Comparison between year 1 (no physiotherapy service) and year 2 (with the physiotherapy service) of the trial group of stores

		Trial group of stores, Year 1	Trial group of stores, Year 2	
Number of full-time employees[1]		1	1	
Paid time – 39 hour week	hr/year	2,028	2,028	
Planned absence	hr/year	156	156	Vacation
Unplanned absence	hr/year	65	50	All unplanned absences including illness
Productive time	hr/year	1,807	1,822	Paid time worked
Wage paid directly to the worker	£/hour	4.60	4.60	
Administrative costs	£/hour	2.20	2.20	£4,462/year
Employment cost	£/hour	6.80	6.80	
Productive employment cost	£/hour	7.63	7.57	Cost per paid hour worked
Overtime (£7.00/hour)	£/year	455	350	Only overtime worked to make up for unplanned absence
Intervention cost per employee	£	–	46	The average cost for the physiotherapy service
Total employment cost	£/year	14,246	14,141	
Savings	**£/year**	–	105	**Overtime cost savings**
Pay-back period	**months**	–	6	

1.   To overcome the complexity of the variations in times worked by the employees, the data have been simplified to show one full-time employee.
     The data for this table are in the productAbilityBasic; File\Examples\Retail – experiment A.bgl).

effect of the service will be diluted by 'contamination' of the data by other reasons for unplanned absence.

The physiotherapy service is a known cost and the cost for an employee to be absent is also known, as the employer pays the employee when they are on illness absence. However, customers still have to be served so that, when an employee is absent, others have to work overtime, which was the case here. In some workplaces extra employees are employed to cover for absences but the net effect is the same; unplanned absences are a cost to the enterprise.

The data that we have used in this case study are the costs of employment with some overheads but, to overcome the problems of varying employment hours (many of the workers are part-time), we have analysed the costs in terms of one full-time employee. In the complete version of productAbility various employee working hours can be accommodated, but this is not possible in the basic version.

The results in Table 5.6 for Experiment A seem insignificant – £105 per full-time employee per year is not a great saving until one expands that to the group of 17 stores, with about 1,600 full-time equivalent employees, and then the figure becomes nearly £170,000 per year. In addition, as the pay-back period indicated in Table 5.6 is six months, the project is worthwhile in financial terms.

Table 5.7 Experiment B. Comparison between the trial group and the control group of stores during year 2

		Control group of stores, Year 2	Trial group of stores, Year 2	
Number of full-time employees[1]		1	1	
Paid time – 39 hour week	hr/year	2,028	2,028	
Planned absence	hr/year	156	156	Vacation
Unplanned absence	hr/year	59	50	All unplanned absences including illness
Productive time	hr/year	1,813	1,822	Paid time worked
Wage paid directly to the worker	£/year	9,329	9,329	
Administrative costs	£/hour	2.20	2.20	£4,462/year
Employment cost	£/hour	6.80	6.80	
Productive employment cost	£/hour	7.61	7.57	Cost per paid hour worked
Overtime (£7.00/hour)	£/year	413	350	Only overtime worked to make up for unplanned absence
Intervention cost per employee	£/year	–	46	The average cost for the physiotherapy service
Total employment cost	£/year	14,204	14,141	
Savings	£/year		63	**Overtime cost savings**
Pay-back (months)			9	

1. To overcome the complexity of the variations in times worked by the employees, the data have been simplified to show one full-time employee.
   The data for this table are in the productAbilityBasic; File\Examples\Retail – experiment B.bgl).

Experiment B (Table 5.7) compares the control group of stores with the trial group for the same year, the year of the introduction of the physiotherapy service. The results when analysed in this way seem to be less encouraging than for Experiment A; the yearly savings for a full-time equivalent employee are only £63 and the pay back-period is nine months.

A summary of the cost-benefit analysis for the introduction of the physiotherapy service is shown in Table 5.8.

Of course, one could always argue whether these calculations are logical and the results valid, as so many other factors are not included in the calculations, for example:

- trading conditions vary from one year to the next
- variation in illness patterns over time
- differences in socio-economic structure of the customer base between the stores
- relationships between staff and management.

Thus it is imperative that one looks at trends. From the summary in Table 5.8, the effect of the introduction of the physiotherapy service shows a pay-back period of six months by one method and nine months by another method; by whichever

Table 5.8 Summary of the cost-benefit of the introduction of the physiotherapy service

	Year 1	*Year 2*	*Comments*
Experiment A	Trial stores with absence reporting system	Trial stores with absence reporting system and physiotherapy service	Savings of £105 per full-time employee per year due to the physiotherapy service. Expanded to the group of 17 stores; nearly £170 000 savings per year and pay-back period of 6 months
	Year 2	*Year 2*	*Comments*
Experiment B	Control stores with absence reporting system	Trial stores with absence reporting system and physiotherapy service	Savings of £63 per full-time employee per year due to the physiotherapy service. Expanded to the group of 17 stores; about £100 000 savings per year and pay-back period 9 months

interpretation the physiotherapy service pays for itself and shows a profit for the retail stores. Due to the complexity of the experimental system, as well as the simplifications we have used in the calculations, it would take a very brave researcher to stand by the absoluteness of either of these figures, but there can be no doubt of the general effectiveness of the physiotherapy service.

Clearly, whichever way one looks at the results one can be reasonably sure that the physiotherapy service had a beneficial effect on reducing absence rates and that the reduction was cost-effective from the employer's point of view.

In Chapter 2 we show that economists simplify their data (assumptions) but that their system is often so complex that you can be led astray by not realising what and where the assumptions are. That we had to simplify the data in this case to get a clear answer is not necessarily a bad thing; as long as you know where the simplifications occur and why, then you are not likely to be led astray by the results.

There is always the problem of the Hawthorne effect, and that will be the case no matter what new measures are introduced. Part of the Hawthorne effect has been nullified by not using the first four months of data (absence rates) after the introduction of either the absence reporting system or the physiotherapy service, but one can only see the long-term effectiveness of new systems after some time (perhaps years) has passed; unfortunately by then many other confounding factors will be introduced.

This analysis supports the 'feelings' of the store managers regarding the positive effects of the physiotherapy service. Up to the end of year 2 the cost of the physiotherapy service was borne by the Risk Management and Insurance Department but, at the conclusion of the experimental period, the Insurance Department withdrew its support; without exception the 17 managers in the trial group of stores decided to keep the physiotherapy service as a cost against their own budgets.

One of the fascinations, and certainly intellectual challenges, of workplace analysis is this uncertainty and having to interpolate between the various factors. Nevertheless, such workplace experiments must be done as they cannot be made other than in the real-life situation.

The Hawthorne effect

In the 1920s and 1930s Elton Mayo and his co-workers were studying the effects on productivity at the Hawthorne Works of the Western Electric Company (USA) of variations in the work conditions. These variations included changes to the lighting levels, pay rates, hours of work and rest breaks. No matter whether the lighting, for example, was increased or reduced, the productivity went up. The conclusion was that the increased morale of the workers, due to their selection as a special group and the presence of the research workers, gave the productivity effect. On removing the stimulus of being a special group the productivity eventually dropped back.

Thus the 'Hawthorne effect' has become the name given to a change in working conditions or working environment that has an initial effect but does not last after the newness or novelty has worn off.

5D.3 A change of work organisation – a comparison between two factories

There has been much discussion about the advantages and disadvantages of the traditional Taylorist/assembly line system and a multi-skilled workforce. Major advantages of the first are that the workforce is cheaper to employ and requires minimal training; major advantages of the second are that, due to extensive training, unplanned absences have less effect on production and the workforce, by having more interesting work, will be more stable.

An engineering employers' organisation decided to investigate the matter by comparing two factories manufacturing similar products. The two factories were similar with regard to type of product and the size of the production departments as well as the socio-economic environment of the towns in which the factories were located.

The investigated differences between the two factories were in the work organisation:

- factory 1 worked according to traditional Taylorist principles, whereby each job is reduced to its simplest components with very short cycle times
- factory 2 used multi-skilling with longer cycle times, having changed its production system from a traditional assembly line two years before this investigation was made.

Each factory employed about 200 people. The comparison was made between a department in each plant, each department having about 50 people whose work was the assembly of electro-mechanical parts (Table 5.9).

The employees in each factory were paid comparable wages and the on-costs for workers' compensation, taxes and supervision were similar.

Taking into account overtime, training (related to turnover rate) and productivity (measured as the reject rate):

Table 5.9 Some information regarding the two departments in the two factories

	Factory 1; Taylorist assembly	Factory 2; multi-skilling
Number of employees	46	52
Average age of employees (90% female/10% male)	25	32
Work cycle time (minutes)	<0.5	3 to 10 (average 5)
Sickness absence (hours/employee/year)	392	264
Overtime (hours/employee/year)	68	115
Labour turnover (% per annum)	50	13
Product rejects by customers (%)	2.7	0.0

- factory 1 – productive employment cost 242 SEK/employee/hour
- factory 2 – productive employment cost 223 SEK/employee/hour.

a reduction in employment costs for factory 2 over factory 1 of about 8%.

It was not possible to compare the unit cost of the product as the two products were not identical, so the only point of comparison was the productive employment cost.

This seems to be a clear-cut case of the value of multi-skilling over Taylorism, and we could leave it at that and give employers who use multi-skilling a 'nice warm feeling' that they are doing good for their employees (lowered injury absence) and to themselves (lower costs). However, the experiment did not turn out quite so cleanly.

There are, of course, difficulties in making comparisons between companies. Great trouble was taken to match these factories for the type of assembly work, size of enterprise, socio-economic factors and so on, but the match was not, and can never be, perfect. There is a time factor involved, so even if various factors are similar at the start of an experiment they may not be similar at the end, a year or so later. Some of the confounding factors identified during the course of the experiment that were not identified *before*, or had changed *during*, the experiment were:

- differences in employment opportunities in each town
- differences in the relationship between company management and their head offices
- differences in complexity of manufacture and product
- differences in standards acceptable to the customers.

There was an important difference in the relationship between factory middle management and their head offices. In one company the production schedule was arranged locally and in the other it was determined by head office (situated in another town) the night before! The latter situation led to difficulties in arranging production schedules and had a serious effect on productivity.

The local situation of unemployment, different in the two towns, had clear effects on the labour turnover rate and was another confounding factor.

Although the experimenters came to the conclusion that multi-skilling was more cost-effective then the older Taylorist methods, there was a grey area of doubt.

The greater the complexity of the experiment, the less can we be sure of its interpretation.

These studies illustrate some of the factors that should be measured before and after a large-scale intervention to be able to interpret the effectiveness of the intervention. However, they also show how unexpected factors confound the interpretation of the experiment.

5E Manual handling

5E.1 Summary

Many ergonomics interventions are directed at preventing musculoskeletal injuries in the workplace. Back, arm and shoulder injuries are still prevalent and a cause of much pain, suffering and cost to the injured workers, cost to the enterprises that may have to pay the workers even when they are not at work, and cost to the society that may have to pay for medical treatment and pensions in extreme, but unfortunately not rare, cases of serious and long-term injury.

Even with considerable mechanical equipment available there are still many tasks that either cannot be mechanised or that seemingly are too expensive to be mechanised.

These next three cases are illustrations of ergonomics interventions which, in all cases, had the sole aim of preventing musculoskeletal injuries in manual workers. In all three cases, as well as success in injury prevention, there was an unexpected financial benefit to the enterprises.

The cases illustrate three forms of intervention:

- small, almost no-cost changes to the work methods where the management and workers experimented to find the best solution (section 5E.2),

- the traditional ergonomics intervention where an ergonomist altered the working heights to overcome poor posture (section 5E.3)
- large capital investment where engineers designed new equipment for injury prevention (section 5E.4).

5E.2 Warehouse work – truck loading

The manager of a warehousing and transport enterprise became concerned when a number of his warehouse staff and drivers began reporting neck and shoulder pain. When the management team did a risk assessment, they identified the source of the shoulder and neck pains as arising from loading particularly bulky packages into the delivery trucks. The enterprise had a long-term contract to deliver this particular product and, on an average day, ten trucks of this product would be loaded for delivery to the various customers.

The product was packed into individual packages, each about 1.5 m wide and 0.4 m in diameter and weighing 8 to 10 kg. The packages were wrapped in a slippery plastic and, with no handles and the contents being soft and flexible, the packages were awkward to handle.

The management team assessed that loading the packages into the trucks above the packers' shoulder and head heights was the major problem. The weight of individual packages was not high but their soft, flexible character made the packages awkward to lift up and, as the rows filled, it required pressure to push the last packages into position on each row (Figure 5.19).

Figure 5.19 Stretching well above head height to load the packages led to shoulder and neck discomfort.

Figure 5.20 Use of a simple platform enables the men to work safely and allows the truck to be loaded more fully.

Table 5.10 Cost-benefit analysis for loading and delivering the bulk packages

	Initial case	Improved loading
Employees	3	3
Full-time equivalent (FTE) (40 hour week)	2.2	2.5
Wages (units[1]/hour)	10.00	10.00
Time taken to load one vehicle (minutes)[2]	35	45
Time taken to load the entire fleet (hours/week)	29	34
Number of vehicles loaded per day	10	9
Total employment costs for the three men (units/year)	50,220	55,690
Reduced productivity (%)[3]	11	5
Cost of reduced productivity (units/year)	4,975	2,650
Truck costs. Includes driver and fuel (units/truck day)	218	218
Truck costs. Assumes a 46 week year (units/year)[4]	501,400	451,260
Total costs (units/year)	551,620	507,000
Intervention costs. Management and warehouse staff time (units)	–	625
Savings (units/year)	–	**44,700**
Pay-back period	–	**Less than one month**

1. For reasons of confidentiality, proportional 'units' are used for costing.
2. The 'improved loading' situation took ten minutes longer to load each truck than in the 'initial case'. Although there was job rotation, for the purposes of calculation it is assumed that it took three men 29 hours (initial case) or 34 hours (improved loading) per week to load the trucks.
3. Unfilled capacity of the vehicles after loading.
4. In the initial case on average ten trucks were loaded and in the improved loading case nine trucks were loaded each day. The truck running costs are added to the equipment costs in the allocated costs screen. The data for this table is in the productAbilityBasic; File\Examples\Manual handling – truck packing.bgl.

The drivers and management team 'brainstormed' several solutions. The enterprise had no control over the way the product was packaged by the producer, so solutions had to be feasible at the local level. The solution chosen was to load rows of product at the front of the vehicle to shoulder height and then pack two layers along the floor behind these rows. A layer of boards was placed on these rows and one man climbed up and stood on the boards; the individual packages were passed up to him and he could then pack the top rows of the truck from this raised platform (Figure 5.20). This stable platform enabled the warehouse staff to work at a height that gave them better mechanical efficiency and eliminated most of the loading above head height.

The new packing system added ten minutes to the loading time required for each truck but, despite this extra cost, it was decided to implement the system. In safety terms the new system has been successful as the warehouse staff have found the method more comfortable and they are no longer getting neck and shoulder pains.

Over the following few months the warehousing and transport enterprise's manager noticed that the warehouse staff were consistently loading more product per truck. He observed that, because they were now packing the top layers of the truck from the platform, the warehouse staff were able to fill the top layers better. Using the original loading system each truck was packed to 89% volume capacity but, with the new system, the capacity was increased to 95%. This made such a difference that the drivers were able to load the entire day's deliveries of product into one fewer truck.

Table 5.10 shows the cost-benefit analysis of the improved loading system. For reasons of confidentiality proportional 'units' are used for costing. In this analysis only the direct wage costs, employee productivity and truck running costs are used. Due to the extra time taken to load each truck, from 35 minutes in the initial case to 45 minutes after the improved loading technique, extra employment costs were involved.

Based on the improved loading, which gave an increased productivity for the drivers and a reduction of one truck required each day, and despite the increased employment costs, there was a net saving of about 45,000 'units' per year. The pay-back period was less than one month. As the pay-back period is so short, there would be little point in deriving all possible costs as their additional effect would be marginal.

There can be an argument that, in this particular case, the reduced productivity (which is a cost against employment) is included in the truck costs (improved loading leading to reduced truck costs). If reduced productivity is taken out of the equation, the net effect is to reduce the initial case total costs to 546,600 units and the improved loading total costs to 504,300 units, leading to a saving of approximately 42,000 'units' per year. The pay-back period is still less than one month.

5E.3 Manual handling made easy: barrel handling

This case study illustrates a traditional, although excellent, manual handling ergonomics intervention designed to prevent musculoskeletal injuries as well as showing increased productivity with lower unit cost. It also illustrates the way in which a cost-benefit analysis can be used as a multiple step process from reduction of injury, reduction of maintenance to an increase in output.

Working as a labourer in a brewery is considered to be a job for the strong; manipulating full beer kegs, that weigh 100 kg or more, needs strength as well as good handling technique. However, strength is not a complete protection against, for example, a back injury when the working conditions are not suitable, as was the situation here.

The work station was the wash line, where empty aluminium beer casks were received from the public houses and washed ready for re-use. Preparatory to washing each barrel the plastic keystones, which are the top caps, were removed by levering out with a chisel and the shives, a wooden bung on the side of the barrel, removed with a hammer and chisel.

The enterprise's occupational health and safety advisers had inspected the wash area and had determined that the work system, and consequent injuries to the shoulders and backs, were not acceptable and that the system had to be made safe. Job rotation, although practised, was not effective in reducing back injuries as other work tasks included bending.

At about the same time as the occupational health advisers' visit, management had decided to increase the line's output but realised that the injury rate would almost certainly increase too.

Through workplace meetings management involved not only the engineers but the operators in finding a solution to prevent injuries and increase output. Although many suggestions were made, it was realised that there was not enough experience within the group to reduce the injury rates and, at the same time, increase production. The organisation's ergonomist was called in to advise.

The original work methods

The casks were mostly 11 and 22 gallon capacity, with occasional 36 gallon casks (50, 100 and 164 litres respectively) and the weights of the empty casks were 9, 16

Figure 5.21 To remove the plastic keystones from the tops of the beer casks required forceful arm movements leading to shoulder injuries.

and 28 kg respectively. The wooden shives were about halfway down the sides of the casks and approximately 26 cm, 31 cm and 37 cm above the base of the casks; that is, at or below knee height even for the smallest of the operators.

Before the casks can be washed, the plastic keystones at the top of the casks and the wooden shives at the side of the casks have to be removed manually. The casks were presented to the operator on a floor-level conveyor belt; the casks were lifted or slid from the conveyor belt to the floor and a chisel used to lever off the plastic keystone (Figure 5.21). The chisel was then inserted into the side of the shive and hit with a hammer and, after several blows, the wooden shive would either be levered off or disintegrate and fall out (Figure 5.22).

The force and repetition needed to break and remove the wooden shives using the hammer and chisel led to musculoskeletal shoulder injuries. Back injuries were caused by the repetitive manual handling of the casks off the conveyor and having to bend to strike the shives. As the platform was at floor level all the operators had to bend their backs, as shown in Figures 5.21 and 5.22.

When the shives were removed the wooden pieces fell to the floor, forming a tripping hazard as well as getting trapped in the conveyor belt and causing damage to the belt. When the belt was damaged it had to be stopped while a mechanic repaired the damage and this took about 15 minutes for each stoppage; on average, a belt stoppage occurred three times each day.

Reduction of injury

Resolution of the causes of injury required a reduction in:

- the degree of bending (for back injuries)
- manual handling (for back injuries)
- the effort required to remove the shives (for shoulder injuries).

Figure 5.22 To remove the wooden shives from the sides of the beer casks required bending and hammering leading to back and shoulder injuries.

For removing the shives without the men having to bend, the shives should be at about elbow height. The most suitable solution would be an inclined conveyor belt so that the men could position themselves at the most suitable place along the conveyor belt, the actual position being dependent on the elbow height of the individual man and the size of the barrel. The slope of the belt would have to be low so that the casks remained stable, and thus a long belt was needed. However, there was a shortage of space so that using a long inclined conveyor belt was not practicable.

Instead, the ergonomist determined the average elbow height of the men working there and, after allowing for the heights of the casks, determined the most suitable conveyor belt height. Of course, the height of the conveyor belt had to be low enough so that the keystone cap on top of the barrel could be removed. Although a compromise solution, it has been effective in reducing, and in fact eliminating, absence due to bad backs.

There was a two-step solution to reducing shoulder injuries:

- the first and immediate solution was to keep the chisels sharp and use chisels with a rubber grip to reduce shock when the hammer strikes the chisel
- the second step was to replace the chisels with a power tool.

Such solutions are at almost no cost; not more than the cost of a single belt breakdown, for example.

Increase of output

It is frequently the case that, even though the work station has been improved, when output is increased there is the risk of an increase in injuries and this is even more

Table 5.11 The basic data used in the analysis

	Initial case[1]	Improved work case[1]	Comments
Employees in the deshiving work station	2	2	Full-time, 40 hours per week
Paid time (hours/employee/year)	2,080	2,080	52 weeks per year
Paid time absent	216	0	Injury absence
(hours/employee/year)	200	200	Vacation
Productive time		1,880	
(hours/employee/year)	1,664		
Wage (£/employee/hour)	9.62	9.62	Paid to the employee
Other employment costs		3,873	Supervisory and other costs
(£/employee/year)	3,873		
Productive employment cost		12.70	Hourly cost when present at
(£/employee/productive hour)	14.35		workplace
Intervention costs (£)	–	8,830	For capital and management time

1. The *initial case* refers to the original work situation before changes were made; the *improved work case* refers to the work station after the intervention. These are the terms used in the Productivity Assessment Tool.

likely when there has been a compromise solution to a health and safety matter, as was the case here.

It was surprising, as well as pleasing, that although there was an increase in productivity of over 70% (from 210 to 360 shives per hour) there were no further injuries during the following year. The reduction of injury was effective even though a 'not-perfect' solution, according to standard ergonomics procedures, was used.

As a general rule, all solutions to safety problems need to be followed up after instigation to ensure that the change has been effective. In fact, all changes to a work station whether for safety or other reasons need to be followed to ensure that occupational health and safety have not been compromised.

The basic workplace and employment data is shown in Table 5.11.

Reduction in injury absence and overtime

In this case study we will derive the cost-benefit analysis as a series of steps. The first step is the savings in injury absence only. There were 54 days' (432 hours total for the two men on the wash line) absence due to injury per year prior to the intervention, and Table 5.12 illustrates these costs and the savings due to the changes to the conveyor belt height and the method of removing the shives.

The savings and pay-back period of 18 months are reasonable considering the relatively small outlay (intervention costs), but are based solely on the elimination of further injury absence and associated overtime to make up for the time lost due to injury absence. Although not the case here, it is more frequent that there will be some continuing injury absences; musculoskeletal injuries usually do not heal as soon as the source of the injury is removed and their effect, including absences, can be felt for some time afterwards.

Table 5.12 Reduction in injury absence and overtime in the deshiving work station

	Initial case	Improved work case
FTE employees	2	2
Overtime worked by the two men (£/year)	8,680	2,630
Total cost (£/year)	56,400	50,400
Intervention costs (£)	–	8,830
Savings (reduced overtime) (£/year)	–	6,000
Pay-back period (months)	–	18

The data for this table are in the productAbilityBasic; File\Examples\Brewery – injury absence only.bgl.

It is often the case that reduction in injury absence costs alone does not form a strong enough economic argument for an ergonomics intervention and, for this reason, one may need to look at other aspects of productivity. Although in this brewery the deshiving area was to be improved for occupational health and safety reasons, an 18 month pay-back period may not have been regarded as a good enough return on investment. In this event further data can be collected to 'improve' the analysis.

Reduction in injury absence, overtime and conveyor belt breakdowns

When the cost of the conveyor belt breakdowns is added to the cost of injury absences, the improvements to the work station look even more financially promising. During the belt breakdowns not only were the two deshiving men idle but so was the entire wash line of another nine men. According to the economic model used here, the Productivity Assessment Tool, the *productive employment cost* of the eleven men on the wash line and the mechanic to repair the conveyor belt was £43.05 per breakdown of 15 minutes. This is derived from a productive employment cost of £14.35 per hour (£3.59 per 15 minutes) for the 12 men. By expansion to the full year, at a rate of breakdowns of three per day and taking into account factory closures, the total employment cost of breakdowns was approximately £32,000.

In the Productivity Assessment Tool this cost of breakdowns is added to the 'equipment running costs'. There were some breakdowns of the conveyor belt after the changes, about 15 per year (£650), and this has been included in the data 'for the 'improved work'.

Some enterprises calculate the cost of employment differently from the Productivity Assessment Tool and it may be more acceptable to management to use their figures. This does not imply that one method is correct and the other is incorrect; it merely means that different assumptions are used.

In this case the enterprise estimated that each breakdown of 15 minutes costs £100, which includes factors other than employment costs. Using these figures would add about £70,000 per year to the cost of the breakdowns rather than £32,000 per year (Table 5.13).

The savings in costs are due to the reduction in injury absence, overtime and conveyor belt breakdowns, and this is indicated by the rapid pay-back period of about three months.

Table 5.13 Reduction in injury absence, overtime and conveyor belt maintenance in the deshiving work station

	Initial case	*Improved work case*
FTE employees	2	2
Overtime worked by the two men (£/year)	8,680	2,630
Cost of breakdowns (wash line idle) (£/year)	32,000	650
Total cost (£/year)	88,400	51,000
Intervention costs (£)	–	8,830
Savings (£/year)	–	37,400
Pay-back period (months)	–	3

The data for this table are in the productAbilityBasic; File\Examples\Brewery – conveyor belt.bgl.

Reduction in injury absence, overtime, conveyor belt breakdowns and increase in output

When one adds the increase in productivity (output) to the reduction in injury and belt breakdown costs, the results of the analysis become impressive. As management wanted to expand output by 71%, and on the assumption that that can be done without injuries to the operators, we can say that the system was under-producing by 42%, as follows.

Increase in output from 210 to 360 shives per hour	$\dfrac{(360 - 210) \times 100}{210} = 71\%$
Under-production (reduced productivity) when deshiving only 210 casks per hour	$\dfrac{(210 - 360) \times 100}{360} = 42\%$

In the Productivity Assessment Tool the supposition is that full (100%) productivity is known. In this particular case study the full productivity is the output after the changes to the deshiving work station (360 casks per hour), and the reduced productivity is the output prior to the changes (210 shives per hour). That is, the *initial* case before the ergonomics intervention had a low output rate, or a reduced productivity, of 42%. In the *improved work* case the output rate is assumed to be maximal with no loss of productivity (100%) (Tables 5.14 and 5.15).

The addition of the increased productivity has reduced the pay-back period to less than two months.

Based on the productive employment cost (reduction of injury absence) and the increased productivity, the unit cost of shive removal has been halved.

In summary, one can say that 'yes, it is good to reduce injury rates' but it may not be sufficient to get a project funded or the resources allocated although, in this particular case study, it was the safety considerations that started the project. The desired increase in output in the wash area was a separate event, but it came along at about the same time as the safety improvements. There can be no doubt that if the work station had not been improved it would have been impossible to increase the output without very serious injury consequences.

Table 5.14 Reduction in injury absence, overtime, conveyor belt maintenance and increased productivity in the deshiving work station

	Initial case	Improved work case
FTE employees	2	2
Overtime worked by the two men (£/year)	8,680	2,630
Cost of breakdowns (wash line idle) (£/year)	32,000	650
Reduced productivity		Full output – 100%
(output (deshiving of casks) per hour)	42% (210 casks/hr)	(360 casks/hour)
Cost of reduced productivity (£/year)	20,053	–
Total cost (£/year)	108,500	51,000
Intervention costs (£)	–	8,830
Savings (£/year)	–	57,500
Pay-back period (months)	–	2

The data for the above table are in the productAbilityBasic; File\Examples\Brewery – increased productivity.bgl.

Table 5.15 Reduction in the unit cost of removing the shives

	Initial case	Improved work case
Productivity (shives/hour)	210	360
Cost per shive (based on productive employment cost for each deshiver and increase in output) (£ per shive)	0.137	0.071

5E.4 Manual handling in coal mines

In longwall excavation as the coal seam is progressively removed the mining machinery has to be moved forward. Although heavy machinery and trucks are used to move the longwall equipment, there are still some tasks requiring manual labour. One of these tasks is moving and re-hanging 11 kV cables which are 300 m in length and supply the longwall machinery.

The longwall machinery is moved forward about 100 m, usually once each week. The cables, which are heavily leaded and armoured, are about 65 mm diameter and weigh 11.8 kg/m. Due to the stiffness of the cable, the men would be holding about 20 m lengths above the ground at any one time (about 240 kg of cable) and it would take three to four hours to put up a 300 m run.

In this coal mine the seams are about 2.2 m thick and the 11 kV cables hang from the roof on hooks about 1.9 m above floor level. Union requirements are for not fewer than six people to fix the cable manually onto the roof hooks; to do this they stand on a flat car so that the level of holding and placing the cable over the hook is about shoulder to head height. Mostly four or five men are holding the cable rather than six, as the miners move in rotation so that the one at the front end of the cable moves to the back end and so on.

Clearly this task is a back and shoulder stress problem potentially leading to injury, and the colliery decided that the only safe way to perform this task was through a mechanical cable handing system.

Figure 5.23 The cable handling trailer consists of a cable drum and a lifting arm. It requires fewer men than the manual method when hanging the cable, is faster to use and enables the men to work safely.

The colliery engineering and safety staff visited several coal mines in the area to see if there were any better systems, and came to the conclusion that the various methods they saw were not good enough. They decided that they would have to design a system themselves.

In collaboration with a company that designs and manufactures mining equipment, a cable handling machine was designed and constructed at a unit cost of $50,000 (Figure 5.23). Using the machine, it now takes three men one hour to put up 300 m of cable safely and with considerably reduced back stress.

Prior to the installation of the cable handling machine there had been about six injuries per year, accounting for 1,200 hours of lost time (on average nearly six weeks' lost time per injured miner). After installation of the machine the injuries had been halved to three per year, with a remarkable reduction in severity; only 22 hours of lost time in each year or about one day lost time per injured miner.

All the underground miners, 80 in total, were trained in the use of the machine. The direct wages of the colliery staff who assisted in the design of the machine and the training costs of the miners (lost time calculated as 2.5 hours' training for each miner) were $15,315.

This is an example of a single task occupying only a small part of the working week, with the tasks not being performed by the same men each time. Table 5.16 shows a simplified calculation in which only the variables of the miners' time (as wages) and intervention costs have been used. The pay-back period is 25 months, which is relatively short for this type of industry, and there may be little to gain from adding extra employment costs; as we noted in Chapter 3 one only needs to collect sufficient data, not *all* possible data.

However, if the pay-back period is seemingly too long then the extra effort to collect more data and to apportion extra costs may well be useful; Table 5.17 shows

Table 5.16 Cost-benefit analysis for the introduction of a cable handling machine

	Manual handling of the cable	Design and use of the cable handling machine
Hours per year required to move the cable	1,092	156
Direct wage cost ($ per miner per hour including production bonuses)	33.50	33.50
Miners' wage cost to move the cable ($ per year)	36,582	5,226
(Machinery)		(50,000)
(Design and training)	–	(15,315)
Intervention costs ($)	–	65,315
Savings in wage costs ($ per year)	–	**31,400**
Pay-back period (months)	–	**25**

Table 5.17 Cost-benefit analysis for the introduction of a cable handling machine including supervisory and administrative costs

	Manual handling of the cable	Design and use of the cable handling machine
Hours per year required to move the cable	1,092	156
Direct wage cost ($ per miner per hour including production bonuses)	33.50	33.50
Miners' wage cost to move the cable ($ per year)	36,582	5,226
Supervisory costs ($ per miner per hour)[1]	6.26	6.26
Supervisory costs for total hours on cable handling ($ per year)[1]	6,836	977
Administrative costs ($ per miner per hour)[2]	19.78	19.78
Administrative costs for total hours on cable handling ($ per year)[2]	21,600	3,086
Labour costs for total hours on cable handling ($ per year)	65,018	9,289
(Machinery)		(50,000)
(Design and training)	–	(15,315)
Intervention costs ($)	–	65,315
Savings in labour costs ($ per year)	–	**55,700**
Pay-back period (months)	–	**14**

1. The wage cost for direct supervision (the deputies) apportioned across all the underground miners gives an hourly supervisory cost of $6.26 per hour for each miner.
2. The wage cost for the administrative and other surface employees gives the hourly administrative wage cost apportioned across all the underground miners of $19.78 for each miner.
    The data for the above table is in the productAbilityBasic; File\Examples\manual handling – colliery.bgl.

some of the extra data that can be collected and the effect on savings and pay-back period. In this case we have added the wage costs for the entire colliery as a cost against the underground miners, on the principle that the underground miners are the actual producers of 'wealth'. Depending on the business concepts of the enterprise, to add the wage costs of those 'above' the level of the producers to the cost of employment may or may not be acceptable.

Supervisory costs as a proportion of miners' costs were calculated from the number of supervisors (13 deputies) multiplied by their hourly wage and divided by the number of underground miners (80).

In a similar fashion the administrative costs included the wage costs of all surface workers apportioned across the underground miners on an hourly basis.

Table 5.17 shows the extra *wage* costs for the entire colliery as a cost against the underground miners. The extra costs have the effect of increasing the yearly savings and decreasing the pay-back period.

There are other costs that could be added, for example:

- increased productivity; the reduction of time spent on cable handling means that these extra hours could be spent on other tasks
- reduced time lost through injury; either extra miners were employed or, more probably, overtime was worked
- reduced workers' compensation premiums and medical costs.

To take into account these other costs would also have had the effect of increasing the apparent savings and decreasing the pay-back period. As one only needs sufficient data to support one's case, would it be worth the extra effort to collect these figures?

Reference

Mitchell, E.C., 1997. Do Ergonomists believe in Ergonomics? In S.A. Robertson (ed.), *Proceedings of the Annual Conference of the Ergonomics Society* (London: Taylor & Francis), 45–50.

5F Personal protective clothing and equipment

Contents

5F.1 Summary

There is much contention as to whether or not personal protection safety equipment should be used: does it not encourage carelessness and give a false sense of security? Often personal protective clothing and equipment are taken as the first step in protection rather than the last resort, and used as an excuse for not changing the workplace to ensure a safer environment. It is frequently observed, for example, that ear muffs or ear plugs will be widely distributed in a noisy factory being, so they say, a less expensive method to protect hearing than quietening down noisy machinery.

However, in many cases personal protective equipment is necessary and cannot be replaced by engineering methods alone. One can think of the chemical industry, where impervious overalls are used just in case anything goes wrong, and in metal working (turners and fitters) where skin creams are used to protect against dermatitis; in welding it is unavoidable that the welder looks directly at the weld and the arc (flame) and requires eye protection. Safety boots and hard hats are routinely worn as an added protection in mines and large building sites, for example.

The use of gloves and the use of welding helmets are examples in this chapter which are consistent with the following criteria:

> First and foremost, the selection and proper use of protective clothing should be based on an assessment of the hazards involved in the task for which the protection is required. In light of the assessment, an accurate definition of the performance requirements and the ergonomic constraints of the job can be

determined. Finally, a selection that balances worker protection, ease of use and cost can be made (ILO, 1998).

5F.2 Hand protection and cost control

When ordering a solvent, or other hazardous substance, the workplace supervisor should obtain the relevant health and safety information, usually in the form of material safety data sheets (MSDS). The supervisor's ability to ensure that the correct protective equipment is chosen is made difficult by the increasing number, variety and complexity of hazardous materials available. The complexity of the safety information can lead to errors in its interpretation and the right protective equipment not be chosen.

As well as safety concerns, there are extra costs if a large variety of protective equipment is held in store. In addition, as personal protective equipment may reduce productivity, one has to ensure that a person is not needlessly 'over-protected'; e.g. gloves being thicker or bulkier than is needed for adequate protection.

In a paint spraying area of a motor manufacturer, five basic solvents were used but in many different formulations. Eighteen different types of protective gloves were kept in the company's store but, with such complexity of formulations, the risk of a wrong choice is relatively high. A simplified system was needed to ensure that workers selected the correct gloves when using any particular solvent formulation.

Reducing the number of solvents would improve the situation by simplifying the protection needed but, after a review, it did not prove feasible to reduce the number of solvents used although it was possible to reduce the number of formulations.

The course taken to reduce the complexity of choosing the correct glove was to re-examine the various gloves' effectiveness. The gloves held in store were checked against a manufacturer's glove degradation chart and, from this, the most generally suitable gloves were selected to undergo the 'udder' test. In this test the gloves were turned inside out, filled with solvent and left until they leaked. During this period the gloves were regularly inspected to note the effects of the solvents and, subsequently, the best gloves were selected for the various solvents.

After these tests it was found practicable to reduce the gloves in store from 18 types to only three types. The formulations were then reclassified by their constituent solvent(s), and each formulation was matched to the appropriate glove.

The benefits in safety and cost control were:

- by simplification of the solvent/glove match the chance for error and potential injury was reduced
- by reducing the stocks of glove types from 18 to three, there was an increase in the usage rate of the remaining gloves
- by buying more frequently it was feasible to reduce the overall stock of gloves held in store.

As well as an increase in safety the reduction in the stock of protective gloves held in store reduced the overall cost by 20% per year.

5F.3 High-cost eye protection for welders

In the conventional welding helmet the eye protection filter is very dark and is opaque in normal lighting. Because of this welders frequently have to raise their helmets to see around them and to identify the actual parts to be welded; as both hands may be holding the welding gun when they strike the arc, they have to flick the helmet back into position with a rapid nod of the head – this latter action has been said to lead to neck injuries in welders.

The needs of welders for improved helmets was recognised some 30 years ago when a young doctorate student, working in the Swedish dockyards, saw the potential for improving the conventional, but cumbersome, welding helmets with better ones. At about this time LCD (liquid crystal) displays were coming onto the market, and his idea was to substitute the welders' tinted glass with an LCD display. When lit, the light intensity of the arc activates a photoelectric cell which responds, within a fraction of a second, by darkening the LCD display and protects the welder's eyes.

Needless to say, any new idea takes a while to catch on and, as well as the lack of standards by which to judge the effectiveness and safety of the LCD display, there was the additional cost. An automatically darkening welding helmet costs ten times the cost of the conventional welding helmet, and it is little wonder that it took some time for the product to be accepted in substantial numbers.

Not only is eye protection important for welders; protection against fumes is important too. The enclosure of the face and neck to reduce exposure to welding fumes leads to an accumulation of re-breathed air, with an increase in carbon dioxide in the inspired (breathed in) air: an increase in carbon dioxide is associated with the physiological effect of dizziness leading to breathing difficulties in extreme cases. One can use ducted air (an active safety system) to the welding helmet, but this is expensive and rather more cumbersome than a 'stand-alone' welding helmet. Correct design of the shape of the helmet around the neck will deflect the welding fumes, and improved design of the inside of the helmet will improve air flow to ensure that carbon dioxide does not accumulate (a passive safety system).

In the following case studies a particular brand of automatic darkening welding helmet (Speedglas, manufactured by Hönell International AB, 78041 Gagnef, Sweden) is examined. In these helmets the response to the increase in light intensity is about one millisecond for the LCD display to darken and fully protect the welder's eyes. As well as the automatic darkening glass, the welding helmets have a passive air flow system that effectively reduces the carbon dioxide level in the breathing zone of the welder.

The excellent level of ergonomics design cannot be denied, but is good ergonomics enough to overcome a ten-fold increase in cost over that of a conventional helmet? The following three case studies look at ergonomics versus economics for welders.

Metal bed manufacture

This case study concerns a small company manufacturing metal beds, bed heads and accessories. It is a highly competitive industry and, although this company survived in a market where most of its competitors have gone out of business, it is now facing stiff import competition. High productivity is essential if it is to stay in business.

Of the 16 full-time employees, two are the company owners (father and son) and three are welders. The other employees are concerned with painting, assembly, warehousing and delivery.

A considerable number of spot welds is required in the fabrication of the bed heads and other metal furniture. Each weld has to be a good one or else the weld would

Table 5.18 Cost-benefit analysis for welders – metal bed head manufacture

	Conventional welding helmet	Automatic darkening helmet
Number of welders	4	3
Paid time per worker (40-hour week) (hour/year)	2,080	2,080
Time paid but not worked (vacation, illness, etc.) (hour/year)	256	256
Productive time (actual time worked) (hours/year)	1,824	1,824
Wage paid directly to each welder ($/hour)	17.50	17.50
Employment cost (wages plus overheads; workers compensation, pension and supervisory costs) ($/hour)	21.00	21.00
Productive employment cost (employment cost per productive hour) ($/hour)	23.95	23.95
Employment cost for the welders ($/year)	174,700	131,000
Intervention (purchase of three automatic darkening helmets) ($)	–	2,800
Savings (assumed to occur in one year) ($/year)	–	43,700
Pay-back period (months)	–	<1

The data for this table is in the productAbilityBasic; File\Examples\Welding helmets – bed heads.bgl. The initial case is the conventional welding helmet and the automatic welding helmets are called the 'better helmets' in the case study.

'blow' in the powder colour process (a heat finishing procedure) and would show as poor quality when on final display in a retail showroom – quality is essential.

Due to the large variety of bed head designs and sizes, the output varies considerably on a day-by-day basis. However, ten welds are usually required per bed head and, as the bed head has to be turned around to weld the other side, it means that each bed head requires 20 separate applications of the welding torch. In one hour a welder would weld about twelve bed heads, so that the welding torch is applied to 240 separate welding spots per welder per hour.

Normally the welder would have to raise his conventional helmet for each weld, but the use of the automatic darkening welding helmet has enabled the welder to weld a whole side of the bed head (ten welds) without having to raise his helmet.

In many ways, this is an exceptional case of spot welding as the welders only do spot welds and, except for setting up their jobs in the jigs, have no other tasks (e.g. no long welds).

As this company has been using automatic darkening welding helmets for ten years, they were not able to determine the actual advantage in present-day cost terms over conventional helmets but they were sure that they would have needed more welders. Although the manager and welders independently were sure that the number of welders would need to have been doubled, we have been rather more cautious, in fact highly conservative, and assumed that only one extra welder would be needed.

In Table 5.18 we have indicated that the number of welders would have been four if the conventional helmet had been worn (second column); at present three welders use automatic darkening welding helmets (third column). Although this is a conservative estimate it will illustrate the use of cost-benefit analysis.

A pay-back period of less than one month can leave one in no doubt as to the advantages of using this particular piece of safety equipment, and shows the advantages of

Figure 5.24 A welder spot welding a crane girder wearing the automatic darkening helmet.

exploring the cost-effectiveness of personal protective equipment rather than looking only at the purchase price.

Crane jib manufacture

Cranes and overhead hoists are fabricated in the workshop of a large international heavy machinery manufacturing company. This case study will examine one aspect of their manufacture, namely the welding and assembly of jibs or crane girders about 10 m long.

These girders are of rectangular section requiring angle bars welded to the side plates and then other plates welded on. There is a considerable amount of tacking (spot welds) as well as long sections of weld. With the welder wearing the conventional welding helmet it previously took 90 minutes to spot weld and fabricate the internal parts of the girder; now the welder uses an automatic darkening welding helmet and this time has been reduced to 60 minutes (Figure 5.24).

That this productivity increase is due solely to the automatic darkening welding helmet is reasonably certain; the company purchased these helmets only 18 months prior to the present investigation and there had been no other discernible changes in the intervening period. This is unlike the previous case of the bed manufacturer, which had been using the automatic darkening welding helmets for so long that there were no comparative measurements of the productivity when using conventional welding helmets, one could only make an estimate.

Based on the number of girders fabricated, the reduction of 30 minutes for each of these assembly tasks is a saving of two weeks' work in a year (4%) for this individual welder. On the assumption that the automatic darkening welding helmet is *only* used for girder fabrication, the pay-back period is about two months, as shown in Table 5.19.

Table 5.19 Cost-benefit analysis for a single welder – girder fabrication in crane manufacture

	Conventional welding helmet	Automatic darkening welding helmet
Number of welders	1	1
Paid time per worker (38-hour week) (hour/year)	1,976	1,976
Time paid but not worked (vacation, illness, etc.) (hours/year)	266	266
Productive time (actual time worked) (hours/year)	1,710	1,710
Wage paid directly to each welder ($/hour)	18.00	18.00
Employment cost (wages plus overheads; workers' compensation, pension and supervisory costs) ($/hour)	32.40	32.40
Productive employment cost (employment cost per productive hour) ($/hour)	37.44	37.44
Fixed employment cost for the welder ($/year)	64,020	64,020
Reduced productivity (4%) ($/year)	2,560	–
Total employment cost for the welder ($/year)	66,580	64,020
Intervention costs (cost of helmet)[1] ($)	–	450
Savings (assumed to occur in one year) ($/year)	–	2,560
Pay-back period (months)	–	<3

1. The cost of this helmet is only about half the cost of the helmets in the previous case. This reflects the cost of the different models.
    The data for this table are in the productAbilityBasic; File\Examples\welding helmets – girder fabrication.bgl.

A pay-back period of about eight weeks for the welder is good, but is not as cost-effective if extrapolated to the entire workshop. Even in this particular task of fabricating long girders, the use of the automatic darkening welding helmet on long welds is not necessary as the helmet stays down for long periods. In the welding shop there are eight other welders and, although all now have automatic darkening welding helmets, many of their tasks include long welds rather than spot welds where the helmets are not as cost-effective as the case of the single welder discussed above.

The positive health effect for all welders wearing the automatic darkening welding helmets is that they are less exposed to neck injury, an injury partly caused when the welders, using conventional helmets, nod or flick their heads to put the helmets down. In addition, and due to the design of this particular helmet, there is reduced carbon dioxide in the welders' re-breathed air. In fact, it was the safety officer that introduced these helmets as part of a long-term injury reduction program rather than simply cost-effectiveness on the part of management.

Harbour equipment

This next case study concerns a company manufacturing spreaders used to hoist shipping containers. These are large pieces of equipment which weigh from 12 tons upwards and require a large number of steps in their fabrication.

The time required in the welding process for one spreader is typically 110 man-hours, of which 70 hours (64%) would be spent in actual welding and the balance (36%) in setting up and other non-welding tasks. Assembly and painting are done by other

Figure 5.25 When preparing, or setting up, for welding there is no need to raise the automatic darkening helmet, as the vision through the visor is quite clear.

Figure 5.26 As soon as the arc is lit, the visor darkens.

workers. The company has been using automatic darkening welding helmets for the past ten years and the production manager estimated that there has been an 8% to 10% savings in welders' time compared with conventional helmets. Since using the automatic darkening welding helmets there have been no new cases of neck pains (as a welder of many years' standing, the production manager has suffered with neck pains!) (Figures 5.25, 5.26).

If the saving in welding time is 8%, then as 64% of the welders' time is spent in actually welding, the net saving is about 5% of the welders' time. The 5% reduction in time is used in Table 5.20 to measure the cost-effectiveness of the helmets.

The pay-back period is less than two months (actually six weeks) when substituting automatic darkening helmets for the customary welding helmets and using a gain of 5% in the welders' time.

A 5% reduction in the welders' time is an estimation on the part of the production manager but he is reasonably sure of his figures. If he had been uncertain about the reduction a more conservative estimate could have been used; for example, if the reduction in time is only half the production manager's estimate then the pay-back period is three months – a good return in anybody's terms.

In addition, from the viewpoint of the welders, there is the advantage that the risk of neck injury has been substantially reduced.

5F.4 Cost-effective noise control

Although this section of Chapter 5 deals with personal protective equipment, there are times when improved engineering controls reduce, or even eliminate, the need for such protection. Noise control is one of those instances.

Table 5.20 Cost-benefit analysis for the 16 welders involved in spreader fabrication for harbour equipment

	Conventional welding helmet	Automatic darkening welding helmet
Number of welders	16	16
Paid time per worker (40-hour week) (hours/year)	2,080	2,080
Time paid but not worked (vacation, illness, etc.) (hours/year)	280	280
Productive time (actual time worked) (hours/year)	1,800	1,800
Wage paid directly to each welder (SEK[1]/hour)	112.50	112.50
Employment cost (wages plus overheads; workers' compensation, pension and supervisory costs) (SEK/hour)	193.56	193.56
Productive employment cost (employment cost per productive hour) (SEK/hour)	223.67	223.67
Fixed employment cost for the 16 welders (SEK/year)	6,441,600	6,441,600
Lowered productivity (5%) (SEK/year)	322,080	–
Total employment cost for the welders (SEK/year)	6,763,680	6,441,600
Intervention costs (purchase of 16 helmets) (SEK)	–	40,000
Savings (assumed to occur in one year) (SEK/year)	–	322,080
Pay-back period (months)	–	<2

1.　SEK = Swedish kroner.
　　The data for the above table are in the productAbilityBasic; File\Examples\welding helmets – spreader.bgl.

Noise at this particular light engineering workshop was generated by cutting, drilling, welding and, particularly, pop riveting aluminium sheet metal. The background workshop noise levels were an average of 89 dB(A) with peaks up to 100 dB(A). It is usually considered that noise levels should be below 85 dB(A) to protect against noise-induced hearing loss and that noise peaks add considerably to the risk of hearing loss.

Although the workers were protected from noise-induced hearing loss by wearing ear protection (ear plugs), the combination of noise and ear plugs led to a considerable interruption in communication between workers and supervisors. The company had regarded engineering noise reduction as a financial burden, and the workers were discussing how to act, through their union, against the company in order to get the noise problem resolved.

The company had little option but to resolve the problem by engineering methods, and it was decided to use the usual methods of noise control, viz.

- reduction of noise at its source from the large machinery
- sound attenuation of the walls and ceiling to reduce reflected sound.

Although little could be done to reduce the noise generated by the hand tools, the measures taken for noise abatement of the larger machinery and sound attenuation to the walls and ceiling reduced the workshop background noise levels to 75–80 dB(A) (Figure 5.27).

The question remains: were these noise control methods a financial burden on the company?

Figure 5.27 The panelling above the mechanic is sound-proofed. Although he still gets the full amount of noise from his own work tools, there is a considerable reduction in noise from his colleagues.

Table 5.21 Cost-benefit analysis for noise control in a metal fabrication workshop

	Original workshop	*Workshop after noise control*
Number of metal workers	12	12
Paid time per worker (38-hour week) (hours/year)	2,080	2,080
Time paid but not worked (vacation, illness, etc.) (hours/year)	252	252
Productive time (actual time worked) (hours/year)	1,828	1,828
Wage paid directly to each metal worker ($/hour)	15.20	15.20
Employment cost (wages plus overheads; workers' compensation, pension and supervisory costs) ($/hour)	25.08	25.08
Productive employment cost (employment cost per productive hour) ($/hour)	28.54	28.54
Fixed employment cost for the 12 metal workers ($/year)	625,990	625,990
Reduced productivity (1%) ($/year)	6,260	–
Total employment cost for the 12 metal workers ($/year)	632,250	625,990
Intervention costs (cost of noise control equipment) ($)	–	8,270
Savings (assumed to occur in one year) ($/year)	–	6,260
Pay-back period (months)	–	16

The data for the above table are in the productAbilityBasic; File\Examples\metal products – noise control.bgl.

From the communication point of view it was conservatively estimated by the workers and supervisors that the time lost by each worker due to noise interruptions was five minutes per day. As there were twelve workers in this workshop, the time lost was one hour per day or about 1% reduced productivity; see Table 5.21.

As a result of the noise control measures, communication between the workers and supervisors improved and discussion of industrial action to resolve the noise problem lapsed.

An assumption was made in the cost-benefit analysis that the entire loss of time due to noise interruptions would be recovered. Of course, as this is an assumption it cannot rigorously be shown to be valid. Even with these assumptions, the use of cost-benefit analysis is useful to illustrate that even a small improvement has a calculable effect on productivity, leading to a yearly saving of over $6,000 (see Chapter 2.5 for a discussion about assumptions in economics models).

It was stated by management and workers that the workplace is better for the noise reduction measures and, in particular, that communications have improved. Certainly, industrial relations at the plant have improved, which was the original stimulus to action on the part of the employer.

As is often the case, an unexpected bonus was found. The sound attenuation (insulation) of the walls and ceiling reduced heating losses and provided further savings. In this analysis we were not able to include these costs, but the heating savings would have improved the financial returns.

5F.5 Reference

ILO, 1997. *Encylopedia of Occupational Health and Safety*, 4th edn (Geneva: International Labour Office).

5G Prevention and rehabilitation

Contents

5G.1 Summary

The previous case studies in this chapter illustrate workplace interventions to prevent physical injuries, with economic arguments being used to support the intervention. However, the same general principle of intervention before an injury is also used to prevent diseases and is a major plank in public health, although the arguments used to support the intervention, for example to vaccinate all children against diphtheria, are not necessarily economic ones.

The first case study discusses the prevention of the disease influenza in a workplace. Although the disease is not a work injury in the sense that, for example, the musculoskeletal injuries were work injuries in the nursing home (see Chapter 5C), influenza has an effect on the productivity and costs of the enterprise. It was to contain these costs that a general hospital started an immunisation program to reduce the incidence of influenza among its employees.

However, even well-designed work methods and workplaces do have injuries, and rehabilitation is also an important aspect of good employment. The second case study concerns a manufacturing assembly line where redesign specifically to rehabilitate injured employees was profitable; in the third case study it is shown that even the rehabilitation of a single office employee, an accountant in this instance, can be profitable for the employer as well as the employee.

5G.2 Prevention is better than cure

Immunisation against infectious diseases such as polio and diphtheria is now commonplace and has been instrumental in controlling serious diseases. But what about diseases that do not kill or maim? Except during pandemics and in special segments of the population such as elderly people and chronic asthmatics, influenza tends to be a nuisance rather than a serious life-threatening event.

Although influenza is difficult to control, as the strains tend to mutate rather quickly, vaccines have been developed and in many countries immunisation is encouraged in high-risk groups. In many countries vaccination for the elderly not only is encouraged but is given free of charge.

What about the rest of the population – especially the working population? Would they benefit by being immunised? Due to the constantly changing composition of the influenza viruses in circulation, influenza vaccines must be modified each year to match the current strains of the virus, hence the need for annual immunisation. The key to this is the global surveillance program, the central coordinator of which is the World Health Organization Global Surveillance Program which enables that organisation to make recommendations on the composition of the forthcoming year's influenza vaccines. These recommendations are made to manufacturers in February for the vaccines intended for the forthcoming winter in the northern hemisphere, and in September for vaccines to be used in the southern hemisphere. The whole process from identification of a new strain of influenza to first availability of vaccine usually takes about eight months. This is in complete contrast to immunisation required for diseases such as German measles, where the effects of protection last a lifetime.

The usual arguments against mass immunisation of the population against influenza are:

- a requirement for annual immunisation
- the protection not being 100% effective
- the illness is usually mild and of short duration.

However, even though spells of absence from work may be relatively short, influenza has large economic effects. In the United Kingdom, the days of work lost through influenza may be as high as 150 million. If this is the case, what is the financial gain to be made by an individual employer encouraging its employees to be immunised against influenza?

The overall responsibility for public health care in the United Kingdom is borne by the National Health Service with funding from the Department of Health. The health departments have been encouraging hospitals throughout the country to immunise their staff against influenza. Is it cost-effective for an individual hospital to do so? The cost of immunisation is usually borne by individual hospitals and not by the health departments.

One hospital, which had decided to encourage immunisation, looked more closely at the costs and benefits to its staff and to its budget. Immunisation was offered to all members of staff through a publicity campaign. The nurses who were engaged in the immunisation program went to each work area and the staff came for their vaccination without an appointment, saving time for the staff and administration costs for the hospital. After the vaccination the staff would rest for about 10–20 minutes and then return to work. For the purposes of the cost-benefit analysis, we have estimated that, on average, a person would be away from their work for 25 minutes.

The immunisation program was launched in late autumn, before the expected onset of influenza, and data were collected for the following six-month period. Despite this

Table 5.22 Absence and cost data for the immunisation program for the six months from late autumn to spring

	Control – not immunised	Immunised
Number of employees	5,219	683
Total days absent[1]	16,752	1,652
Average days absence per person	3.210	2.419
Reduction in absence (days)	–	0.791
Total number of spells of absence[1]	2,531	397
Spells of absence per person	2.062	1.720
Average duration of each spell (days)	1.557	1.406
Total wage cost for absence[2] (£)	2,247,500	221,600
Cost per person for absence (£)	430.6	324.5
Reduction in absence costs per person (£)	–	106.1
Total savings in wage costs for those immunised (£)	–	72,500
Intervention cost (immunisation program) (£)	–	**12,400**
Pay-back period (months)	–	**2**

1. The days absence are for spells of seven days or less.
2. The absence costs are based on an average wage of £18.13 per hour and a 37 hour week.

being the third year of the program and the program being well advertised, only 12% of the staff volunteered for immunisation.

The determination of whether or not an individual suffered from influenza requires virological studies as the medical absence certificates provided by general practitioners are rarely specific enough to determine such a diagnosis. Under the circumstances it was decided to use only the absence data that were seven days or less per spell, on the assumption that these absence spells would have included all influenza episodes. This, of course, introduces an error in that the effect of the immunisation (reduced absence due to influenza) will be diluted by other absence causes. It is not possible to estimate the degree of error, but the net result of the dilution effect on the cost-benefit calculations will be to *reduce* the apparent benefits of the immunisation program.

Table 5.22 shows the absence data, and from this it can be seen that there was a reduction in average days' absence and the number of spells of absence for the immunised group of employees compared with the non-immunised (control) group.

For those staff immunised, the effect was certainly positive in that, on average, they were less sick, but did the cost of the immunisation program bring a financial benefit to the hospital?

Due to the limitations of the computer and recording systems, it was not possible to use the work categories of the individual staff members and instead the general employment statistics for the hospital have been used; these included the net wage cost, the average hours of work and the staff numbers. On this basis, it was estimated that the average staff salary on an hourly basis was £18.13. This did not include supervision, insurance or any other overheads that we have used in other case studies, and so the apparent effectiveness of the program will be less than if these other factors had been included.

Other errors introduced included variation in actual working hours between the various categories of workers; for example, nurses work 37½ hours per week and administrative and clerical staff work 36 hours per week, and there was also a relatively small variation of uptake of immunisation between departments.

The direct cost for immunisation, including supplies of vaccine, cost for the nurses to visit the various hospital departments and wards (an outside nurse-supply agency was used), travel expenses and 15% added for administration costs, was £6,900. However, to the direct cost of the program must be added the indirect cost of loss of productivity by the staff when attending the immunisation program. This we have estimated as being 25 minutes or £7.55 per person, which is a net productivity loss of £5, 516. The total cost of the program was £12,400.

Table 5.22 is a summary of the essential cost data and indicates that the saving in reduced absence due to immunisation was £72,000 during the six-month period, or a pay-back period of two months.

It is due to potential errors that one needs to make a decision as to whether or not extra detailed information is required. In Chapter 3 we discussed the limitations on the collection of data, and stated that one needs only to collect enough data to support one's argument. One has to determine whether or not it is worth collecting the extra data – as we noted before; it takes 20% of the time to collect 80% of the data and 80% of the time to collect the extra 20% of data. In this case study, the largest source of error (absences) would lead to a conservative estimate of the program's cost-effectiveness and the effect of the smaller errors, in our opinion, did not have a significant effect on the overall cost-effectiveness of the program.

There would seem to be little doubt of the effectiveness of the immunisation program both in terms of the health of the staff and in cost to the hospital. The year reported here was not an influenzia epidemic year, but it is clear that if the hospital can achieve a higher proportion of acceptance by the staff of the immunisation program the savings in a year of low influenza prevalence will be in the order of £500,000; in a year of an influenza epidemic the savings would be far greater. No further detail is required to prove the effectiveness of the imunisation program.

5G.3 Workplace redesign to incorporate rehabilitation

The enterprise in this case study had manufactured the same electrical product for 40 years. The production technology was based on the principle of job simplification (Taylorism) although various elements of job enlargement had been introduced during this period. In an attempt to increase productivity the enterprise was to build new work areas, and decided to follow the principles of job enlargement.

Over a period of years there was an *apparent* slow increase in productivity due, in part, to an increase in work pace and a reduction of rest pauses for the assembly line workers. These changes led to increased stress to the workers and eventually to an increase in musculoskeletal injuries. Workplace changes were small over a long period, so that one cannot depict a single event as increasing stress beyond acceptable limits.

This *apparent* productivity increase had been based on the output per worker when he or she was working, and did not take into account the cost of absences due to injuries and labour turnover. The high injury rate led to serious employment problems

Table 5.23 Cost-benefit analysis of the old system (Taylorism) and the reorganised work (job enlargement) including rehabilitation work areas

	Old system	Reorganised work	
	Full-time assembly	Full-time assembly	Rehabilitation workers
Assembly line employees	10	10	3 (part-time)
Productive hours/employee/year	1,425	1,540	900
Wages ($/employee/hour)	12.00	12.00	12.00
Fixed employment cost ($/year)	322,560	274,560	44,928
Extra employee cost ($/year)[1]	64,512	–	–
Estimated reduced productivity cost ($/year)	16,128	6,864	–
Total employment cost ($/year)	403,200	281,424	44,928
Recruitment cost ($/year)	15,414	7,707	
Insurance premium and social costs: 10% of wages ($/year)	24,960	28,704	
Total cost ($)	443,600	362,800	
Intervention cost ($)	–	32,500	
Savings ($/year)	–	80,800	
Pay-back period (months)		5	

1. In the old system due to the high injury and turnover rate, 20% extra employees (overstaffing) were required.

which included a labour turnover rate of 40% and sick leave (injury and illness) of 18% per annum. Overstaffing to cope with the high turnover and sick leave rates was 20%.

The cost for training new employees was high as it took 11 weeks to train an assembly worker, during which time his or her production output was about 50% of average production.

Taking all employment factors into account, productivity was not good. Table 5.23 shows the estimate for reduced productivity cost was $16,128 per year.

The injuries sustained in the assembly area were mainly musculoskeletal ones to the neck and upper limbs. Over time, improvements had been made to the physical arrangements of the assembly work to reduce musculoskeletal stress, but it had reached a point where no further improvements were reasonable; the only avenue open for improvements was in the organisational style.

The work stations were re-used in the new work areas to reduce capital outlay but substantial alterations were made to the work organisation which included:

• job enlargement with training so that most tasks can be done by all employees
• job rotation
• all training and rehabilitation is done within the group
• there is now no supervisor; the supervisor's function has been absorbed within the group which takes overall responsibility for its production
• clerical work, previously done by additional staff, is now done by the assembly workers.

The total capital cost for the intervention was $64,512 per year.

The original work areas consisted of groups of ten full-time employees but in the new arrangement this was enlarged to 13 places for each work area; ten for full-time employees and three places for employees who were being rehabilitated.

In a typical rehabilitation program an injured employee works for half a shift, has reduced work output per hour and is allowed to take breaks or pauses as needed.

These new assembly lines have been very successful in rehabilitating injured workers as well as reducing the injury rate. The sickness and injury absence rate has dropped from 18% to 12%, the labour turnover rate has halved from 40% to about 20% and overstaffing has been eliminated with a saving of $64,512 per year.

With an on-going saving of $80,000 per year and a pay-back period of five months this new arrangement and illustrates very clearly that a rehabilitation program can be incorporated into the normal production system and be very successful both for the injured workers and for the employer.

5G.4 Rehabilitation of a professional employee

As we have noted elsewhere, there are frequently difficulties in measuring the productivity of office workers and the difficulty is particularly acute for professional employees. One would have thought that measuring the effectiveness of a rehabilitation program would have presented even more difficulties, but this is not necessarily the case as shown in the following example.

The case study illustrates a method to address the productivity issue and the cost savings for a rehabilitation program. The concept of productive time, discussed in Chapter 4.2.3, is used here.

The employee in this case study, an accountant, enjoyed her work and was in good health except for recurring bouts of lower back pain. The accountant did not consider her complaint to be serious as it caused only occasional discomfort sufficient to require short absences from work. However, as time went by the spells of absence became longer and, although her work was not the initial cause of her back pains, her absences started to affect her work.

The accountant worked for a consulting accounting company and each accountant had their 'own' clients; this specialised style of organisation meant that it was not practicable to employ a replacement accountant on a casual, or part-time, basis to cover absences. It was also apparent that some clients were becoming dissatisfied with her absences and there was the possibility of the firm losing some of its clients.

The productivity measurement is based on the accountant's time and all costs reduced to that of her direct hourly wage. Thus the calculations shown here are not in money terms but in time terms. The assumption is that all time spent by the accountant at work is of equal value (this is not strictly true as we all have some days better than others, but, over a reasonable length of time, there is an average output).

From the employer's point of view, and also the clients', the concept of productive time is valid as the time lost by the accountant cannot be recovered and is a net loss of productivity.

Table 5.24 shows the absence situation for the accountant after she had been working for ten years for the consulting firm.

The employer decided to rehabilitate this employee and engaged an ergonomist to advise on changes to the accountant's office. Her work was administrative in style

Table 5.24 Time lost by the accountant due to lower back pain during her tenth year of employment

	Hours of absence
Two periods of absence for a total of four weeks	160
Two visits to a physician during working hours	2
Reduced working capacity due to pain[1]	20
Total hours lost	**182**

1. Conservative estimate of the unproductive time due to back pain when at work and before taking illness absence.

Table 5.25 Work station improvements calculated in terms of the accountant's hourly wage cost

	Equivalent hours in terms of the accountant's wage
Time spent on the work station redesign by the ergonomist	8
Discussions by the accountant, supervisor, company purchasing officer and ergonomist regarding the work station redesign	4
New furniture (cost per year amortised over 5 years)	28
Total equivalent hours	**40**

and required the usual materials – a computer and other writing materials, telephone, bookcases, filing systems, desk, chair, desk lights and visitors' chairs. Due to the office layout, considerable reaching and stretching was required by her, especially when on the telephone to a client.

The reaching required was reduced by changes to the placement of files and other paperwork, and a standing-height desk was provided so that the accountant could stand or sit as required to relieve her back pain.

The equivalent costs of the office redesign, furniture and other factors are shown in Table 5.25.

The accountant attended an occupational rehabilitation course which taught her how to manage her back condition including postural re-education, stretching and strengthening exercises and developing skills in order to minimise her disablement at work and at home.

Table 5.26 shows the time lost by the accountant during her rehabilitation program. Treatment costs were borne by the insurer and are not included as a cost in the table.

In the 18 months subsequent to the work station changes and the rehabilitation program, the accountant had had no further absences from work due to back pain.

Table 5.26 Time lost by the accountant during the rehabilitation program

	Hours absent from work when attending the rehabilitation program
Attendance at an occupational rehabilitation centre. 2.5 days each week for six weeks	120
10 visits to a physiotherapist	20
One visit to a treating physician	1
Total hours for treatment	**141**

The total cost for rehabilitation, treatment and work station improvements was the equivalent of 181 hours of the accountant's time (Tables 5.25 and 5.26) and, as the absence for back pain by the tenth year of employment (pre-intervention) was 182 hours (Table 5.24), the pay-back period was one year.

6 New supervisory systems

Contents

6.1 Introduction

This chapter will examine some of the effects of computer technology, particularly in the service industries, on the health and safety of employees and will relate this to the ergonomics and economics themes of this book. One of these new service industries is the call centre industry, which will be used here as an example of a computer-driven industry.

The case studies described in Chapter 5, on the whole, were from workplaces having the traditional style of supervision, i.e. where a supervisor directly manages several or many persons. Traditionally supervisors only direct the general work of the workers, not usually the details of the work; the supervisor only says 'do such-and-such' and then comes back to check that the work has been done, and done satisfactorily.

New forms of supervision are emerging due to the advent of computers. Computers have allowed the continuous, automatic collection of data and much of the previous role of middle management in supervising the production process has been made redundant. Consequently the number of supervisory positions has been reduced.

Middle and lower management is thus developing a new supporting role. This supporting role is clearly seen in call centres where the supervisor, or team leader, usually has 10 to 15 people under their supervision and where their tasks include training, career coaching, providing information and other supporting functions.

It is not only middle management that has felt the effects of computer technology. Much of the automatic data collected as a necessary part of production (service or manufacture) relates to individual workers, and this has given the employer a degree of knowledge that was not previously present. Management can call up data about individual workers and measure their work in terms of quantity and, sometimes, quality. For example, if supermarket check-out operators (cashiers) are given a scan rate of, say, 1,500 items per hour they can be measured *during their work* to ensure that they are achieving this rate. This is a far more detailed and potentially intrusive supervision than was possible in the traditional industries.

Such detailed personal knowledge gives the employer increased control over the employee; the employer can select employees who are better performers, leading to competition between employees. There may be little wrong with competition *per se* but, in a situation where the employment is by the hour (precarious workers), the competition is between keeping one's job or others keeping theirs.

In this chapter we will look at call centres in relation to the health and safety of the call centre operators. We would expect that the issues discussed would be usefully applied to any industry using computer-based data collection.

Call centres are only one example of an industry utilising automatic data collection; the same concepts are found in word processing, data entry (including cheque and credit card entry), warehouses, check-out cashiers and eventually perhaps those in mobile work places, viz. commercial travellers, sales representatives, tradesmen and carriers by the use of GPS (Global Positioning System). In fact any activity that is computer-based is open to this style of supervision.

6.2 Setting the scene – call centres as the new supervisory system

Call centres give employment to a wide range of people and thus are of social benefit. This is realised in many countries where there are government incentives for companies to open call centres in areas of high unemployment, often in rural areas but not always so. Glasgow, for example, is a major call centre region but in Sweden most call centres are in small towns well away from the major population centres.

About 80% of call centres, or call centre business, are in-house and the balance are outsourced call centres providing a service for other enterprises. Being a growing industry, there is no useful analysis of the actual work tasks nor any actual tally of operators. The number of call centre operators is certainly high and, in some countries in Europe, is given as 2% of the working population.

In the everyday literature of the newspapers and popular media call centres are often given a bad name: 'the current day sweat shops'; 'Taylorism in the office'; 'stress down the line' and so on. How true is all this and, if it is true, how can ergonomics and economics, in the form of cost-benefit analysis, be utilised to improve these workplaces? To understand and address this issue we need a model of the organisational system in a call centre.

6.2.1 A call centre model

As with any industry, call centres are not identical. Taylor saw manufacturing in the early part of the 20th century being performed by people who were not educated and mostly illiterate; each task was reduced to its simplest steps, each work cycle was short, supervision was constant and the workers worked as individuals.

In the latter half of the 20th century, with a workforce that was educated, Deming's concepts were to make use of this education for the benefit of productivity; the tasks could be more complex and the workers able to control quality; there was less frequent supervision and the workers worked as a team.

We will use some of Taylor's and Deming's concepts to derive a model for call centres so that the ergonomics issues, particularly those of health and safety, can be put into perspective.

The call centre industry covers a wide range of services, from the quantity ('Tayloristic') end to the quality ('Deming') end. These are:

- the *quantity* end of the spectrum, implying little training, highly repetitive tasks with little control of the process and the product by the operator, e.g. directory enquiries where an operator takes more than 100 calls per hour
- the *quality* end of the spectrum, implying more training and higher skills with considerable control by the operator of both the process and the product, e.g. computer help desks, social services support services.

In the call centre industry at the *quantity* end of the quality/quantity spectrum, the 'Tayloristic' end, the employer controls both the process and the end product; the words used by the operators may be scripted for them and the accuracy of the end product is measured by the supervisor listening in to some of the calls. Time spent with callers, calls per unit of time, time away from the desk and so on are collected by the automatic data collection system and immediately available to the supervisor. At this end of the spectrum there is low control by the operator of his or her work, little task variation or complexity but high work intensity; when one call is finished another is waiting with very little, if any, wrap-up time (pause between calls). Process data may also be immediately available to all operators, usually by means of a large wall display of calls waiting and waiting times, which can be an added stress factor.

Call centre operators may sit at their work stations for relatively long periods, leading to musculoskeletal discomfort due to static posture, and section 6.4 will discuss the health issues in more detail.

This high degree of control was present in the older style telephone switchboards and exchanges where the supervisor collected operator data manually. The difference now is that data is collected automatically, is more detailed and can be saved for immediate analysis or analysis at any time in the future. This is not to say that these features are always used, but there is the potential for their use and this is reminiscent of the boss in Charlie Chaplin's film *Modern Times*, who not only can view the workers, but is omnipresent via large television screens.

As the call centre industry matures there is an increasing shift away from Taylor's model with, for example, simple enquiries (directory enquiries, taxi bookings) and standard processes of account payments being taken over by voice recognition.

Typical Taylorism

The following is a typical example of Taylorism and comes from the automotive industry. It typifies a task with little variation or complexity but with a high work intensity.

A worker, a woman in her 50s, used a hand cutter to trim the excess tabs from number plate frame pressings. To keep up with the production line she had to make about 14,000 cuts per day, a cycle time of one cut each two seconds; no skill was required, only speed. As the trimmed parts were metal, considerable hand force was needed to operate the hand cutter, Figure 6.1.

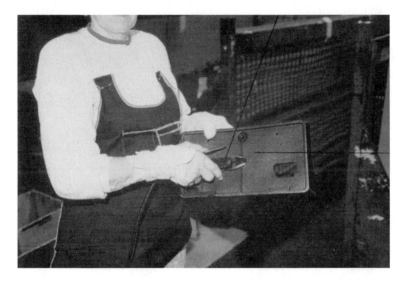

Figure 6.1 The worker had to make 14,000 cuts per shift, so it was not surprising that she ended up with considerable hand and wrist injuries.

The highly repetitive nature of the task and the pressure required for each cut led to very serious hand problems that required many months' absence and expensive hand surgery for the worker (carpal tunnel syndrome). When she came back to work, on the same task, she had her wrists in splints. In this case there was a 'happy' ending as the hand tool was replaced by a small air-powered gate cutter fixed to a bench. The operator needed only to place the part to be trimmed in the cutter's 'mouth' to activate the cut.

However, the change was only to the tools, not to the Tayloristic principle of short cycle times with an unskilled workforce. Although the physical health risks were removed and the woman was literate and could be trained for a more complex task, nevertheless the job was Tayloristic in principle and remained so.

Call centres are labour-intensive and there is a growing trend to offset costs by converting information calls into selling calls – answer the caller's question and then sell them something. Selling in turn requires a different set of interpersonal skills.

This tendency to convert incoming calls into a sales system converts the system, no matter where it is on the quality/quantity spectrum, into a Tayloristic model if the number of items sold (the product) becomes the criterion, not the quality of the process. There are exceptions, of course; government call centres usually have nothing to sell and focus on maintaining a high quality of service.

Six steps to quality

To control the process, with workers being responsible for quality of their output, is a main point made by Deming and taken up by Japanese industry. For example Ishikawa (1985) describes six steps that apply to all levels of employees and contractors, from the general manager to the production line worker:

- determine goals and targets
- determine methods of reaching goals
- engage in education and training
- implement work
- check the effects of implementation
- take appropriate action.

In other words, not only management needs training in *process control* but also the production (or service) workers who have to work within *process control*. It is no longer sufficient to have a quality control officer at the end of the line to reject mistakes; errors are corrected and eliminated within the process. Use is made of the experiences and abilities of the workers at all levels, not solely of management.

At the *quality* end of the call centre spectrum, the 'Deming' end, there is more control of the process by the operators as the incoming callers are varied in their requirements with concomitantly less control by the employer; hence the operators are more highly trained. There is automatic data collection but this has less immediate relevance compared with the quantity-type operations. Operators at this end of the spectrum have more control over their tasks and are able, for example, to take short breaks when needed.

Although we have referred to the two ends of the quality/quantity spectrum in order to approximate the Taylor and Deming models, in fact there is a continuum, illustrated in Figure 6.2. Various aspects of the call centre industry can be placed on a continuum between the extremes, although we have seen no call centre that is to either extreme; each centre has some position along the various continua.

However, Figure 6.2 is by no means the complete picture for call centres. The Taylor and Deming models were developed for manufacturing industry and, although adapted to describe service industries, there are aspects of the service industries that extend beyond these two models. This extension is the direct interaction that operators have with the customers. In manufacturing, and in many service industries, the workers are several steps removed from their customers; neither the production line

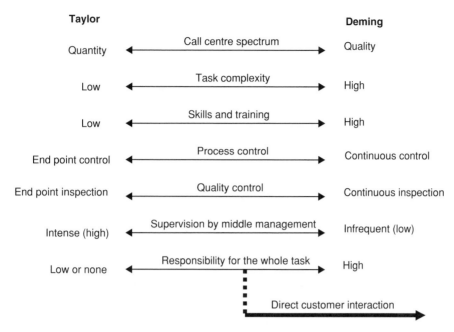

Figure 6.2 A call centre model: the Taylor and Deming concepts applied to call centre operators' tasks.

workers in a car factory nor the kitchen staff in a hotel have direct contact with the customers but in call centres, as with nurses and retail shop assistants, almost the entire working time is in contact with the customers (or patients).

In Figure 6.2 the last continuum (Responsibility for the whole task) has to be extended to the right to include the concepts of customer interaction. Especially in this particular area of customer interaction, considerable stress is, or may be, placed on the operators. Not only do they have the responsibility for the whole task but operators have to be pleasant too; *smiling down the phone* is a term coined to describe this. No matter their mood or the mood of the caller, the operator must be pleasant at all times.

As well as the personality of the operator and the particular task, the degree to which a high level of customer interaction leads to stress depends on several factors under the control of the employer, including support and training.

In the Deming model for the manufacturing industry, workers work as a team and the team members pool their ideas; the team as a whole is acknowledged for their efforts and ideas. In call centres, although the organisational control is through teams, the computer system enables the individual operators to be judged and controlled as individuals, as in Taylor's model.

Where employees are judged as a team there is an incentive to work together, sharing new knowledge and helping each other with their problems. However, where:

- the organisation judges team members as individuals
- supervisors use their power unduly or intrusively

- there is excessive productivity pressure
- there is a down-turn in business
- there is high unemployment

there is more likely to be competition between operators with less sharing of knowledge accompanied by a feeling of 'each one for themselves'. An unfortunate consequence of these factors is that they may lead to stress and ill health of the operators arising from stress.

From Taylorism to multi-skilling

Although not in a service industry, the following case study illustrates the effectiveness of allowing employees responsibility for their work. The workplace employed 16 full-time workers to assemble small electrical components. The product range included about 40 different electrical switches.

Assembly was a batch production process, with work organised along traditional Taylorist lines, viz:

- highly repetitive work with short cycle times
- a lack of variation of the work
- no requirement or scope for individual initiative and responsibility
- little intellectual stimulus
- an absence of scope for personal development.

There was a high injury and absence rate with the injuries, being mostly musculoskeletal to the neck and/or shoulder. Productivity was poor, it took 12 days to fill an order and only 10% of orders were filled in the time required by the customers.

The work stations were altered to avoid musculoskeletal injuries and other changes were made, broadly along the lines advocated by Deming, by broadening the work content and increasing workers' responsibilities.

Job rotation was introduced and all workers trained to do all the jobs from entry of materials, assembly of all types of switches, packaging and dispatching. Three-quarters of the workers chose also to take supervisory roles.

Each worker was given a total of 125 hours' off-the-job training followed by approximately 125 hours' on-the-job training plus a further two to six weeks' training in supervisory tasks.

The previous turnover rate of 39% was reduced to nil in the first year of the changes, and the absence rate dropped from 36% to 30%. Musculoskeletal injuries heal slowly, and a rapid reduction in absence rates is not to be expected.

On the productivity side, over-employment, to make up for the high absence rate, has been reduced, stocks of finished products have been reduced by 67% and 98% of orders are finished on time.

The project met all its health and productivity aims and the pay-back period was one year.

Having set the scene, and acknowledging that the call centre industry has a *potential* for injury to the operators, what can be done and how do we use an economic argument to best advantage?

6.3 Comparison between technologies

To our way of thinking there is a direct comparison between the introduction of new computer technology (word processors, data entry) to offices in the 1970s and 1980s and the introduction of new computer technology (call centres) in the 1990s and 2000s. This comparison can help us to point to the type of intervention needed to control injuries in call centres.

In the early 1980s, in Australia and Sweden and later in other parts of the world, there were many cases of upper limb musculoskeletal injuries in office workers. This was at a time when word processors were replacing typewriters and there was an accompanying change in office organisation. Previously secretaries had several tasks, not solely typing, and this enabled them to be active and not tied to their desks. When word processors were introduced the secretaries, now called word processor operators, had a reduction in task variation and sat for long periods only typing. The modification was from one of varied work tasks to one of a single task, with the organisational change leading to:

- the workers being static, sitting in constrained postures for long periods
- undue productivity pressure being put on the operators
- lack of control by the workers over their work.

We believe that the same situation applies to many call centre operators.

In a study of upper limb musculoskeletal injuries of word processor operators (Oxenburgh, 1984), improving the physical aspects of the work station (desks, chairs, lighting and so on) did not reduce musculoskeletal injuries; the only significant difference between the injured and control workers in this particular office environment was that the incidence of musculoskeletal injuries increased markedly if the word processor operators were at the computer for more than five hours per day.

The injured people were mostly female, and this led to the belief that upper limb musculoskeletal injuries were a function of being female. In fact, they were a function of the workplace structure. In a similar vein, at present younger people tend to occupy operator positions in call centres, and in one study the call centre operators suffered from more musculoskeletal discomfort than the control group of older computer users (Norman *et al.*, 2001). Injury is a function of neither gender nor age, but of the work and workplace organisation.

Organisational psychologists often use a simple but useful concept to illustrate where one could expect adverse health effects. Work *intensity* alone does not usually lead to ill health or injury if the worker has *control* over his or her work. However, if there is both high work intensity and low control over the work then the injury risk increases.

If we look at the organisational aspects of work in the above office example, the intensity of work was high and the word processor operators had little or no control. Some of these same organisational features that were found in office studies may be present in at least some call centres, for example:

- there is lack of control by the operator over both the amount of work and when it appears throughout the day
- the operators sit for long periods in one place, potentially leading to static posture discomfort
- poor knowledge of computer systems
- the hourly work flow can usually be estimated ahead of time from past records and the employer can use this information to reduce the number of operators to an absolute minimum. In a government centre we visited they were able to estimate the work flow with an accuracy of 3% one month ahead
- an additional stressor, previously mentioned, is having no control over the reactions of the callers no matter how unpleasant they may be.

What research work has been published on call centres indicates that all is not well. One study of four in-house call centres included small and medium-sized, government and private sector call centres with the work tasks at about the middle of the quality/ quantity spectrum (ATK, 2001). The methodology of this investigation was by questionnaires of the operators and interviews with management. Survey of the work stations (furniture, lighting, noise and air quality) and software used by the operators was assessed by ergonomists. The positive findings were that:

- there was a high degree of satisfaction by the operators with the work
- the operators had reasonable control of the work and could take breaks as they wanted to
- they identified with the work and were proud of the work and workplace
- the training was sufficient
- the support they received from their colleagues and management was satisfactory
- the labour turnover was low (10%).

All these are indicators of good places to work. However, about 50% of operators experienced eye strain and/or upper limb, shoulder or back discomfort on at least some days during each month.

If these four call centres were seemingly such good places to work, why was there musculoskeletal discomfort? Although the work was satisfactory from the operators' psychological viewpoints, it was the physical aspects of constrained posture over long periods that gave rise to the musculoskeletal discomfort and possibly the eye fatigue too.

There have been academic studies of call centres (Taylor and Bain, 1999; Norman *et al.*, 2001) and studies by some unions representing call centre operators and Occupational Health Authorities (Health and Safety Executive, 2001) The latter were mostly based on the results of questionnaires sent and/or returned by post or e-mail, although the percentage of replies was often low. Nevertheless, these studies all tended to show that there are musculoskeletal injuries, eye strain (visual fatigue) and stress problems, voice and possibly hearing difficulties among call centre operators.

Partly based on present research of call centres and partly based on experience with previously introduced computer-based technology, in the next section we will discuss potential ill health effects for call centre operators and methods to reduce these effects.

6.4 What can be done to ensure good health among operators?

Derived from the concepts of Taylor and Deming, in this chapter we have started to evolve a model for call centres which included arranging call centres along a spectrum. We have also examined the introduction of another computer-based technology (word processors) to see what can be learnt from the past: the historical perspective. Now we will see what are the potential or actual sources of injury in call centres and how they can be addressed; Figure 6.2 may assist in stimulating thoughts on this.

On the assumption that at least some call centres may give rise to injuries, what can be done? This section is not intended to be an exhaustive examination of the causes and cures of injuries but a signpost to stimulate thought in that direction, and the case studies in section 6.5 indicate how financial considerations can assist in implementing an ergonomics intervention. It must be remembered that, most commonly, several effects will occur at the same time and separating them is a convenient analytical method only.

6.4.1 Workplace and work station design

Standards for work station design, including furniture, lighting, noise and air quality, for a call centre are no different from those of other computer work stations. Most countries now have design rules, guidance notes or standards on office and computer work station design, chairs, tables, lighting, air quality, noise, screen legibility and so on, and these should be consulted. Other areas of information are from unions as well as the standard ergonomics literature.

Unless there is a serious mismatch between the individual operators and the furniture, furniture *per se* is not a major contributor to adverse health effects; rather it is the general work station layout that restricts movement of the operators leading to static posture, discomfort and eventual injury.

Good furniture design can assist in preventing static posture. The concept of a standing/sitting desk, to ensure that the operators are more active and change posture frequently, can be used as a preventive measure and was the aim of a previous study on data entry operators (Winkle and Oxenburgh, 1990). As the operators moved about more (were less static) there was a decrease in perceived discomfort about the neck and shoulders without there being a significant decrease in productivity (keystroke rate per hour).

The position of desk equipment can be important; if the screen is not at the optimal visual distance (horizontal plane) or is too high or too low (vertical plane) for the individual operator it can lead to eye strain and/or neck discomfort. Similarly, the keyboard and mouse must be placed where they will cause least awkward shoulder posture, although in call centres the keyboard is usually only a minor part of the task. The incorrect placement of the mouse and mouse pad have given rise to many shoulder injuries, and constant attention should be given to the use and placement of the mouse. It is most suitably placed close to the centre line of the body with the elbow about right angles and shoulder relaxed.

There must be sufficient room for any papers, books and writing material on the desk and our personal preference is for a flat desk, easily height adjustable and with sufficient depth.

Chairs should be adjustable and meet the relevant standards of design and ease of adjustment.

Knowledge about the optimal placement of desk equipment and adjustment of furniture can be incorporated into the operators' training, with the team leaders ensuring that each operator adjusts their work station for optimal comfort and safety. If 'hot desking', time must be allowed at the start of the shift for each operator to adjust the work station layout and furniture.

Average noise levels are well below the level that will lead to hearing loss, but headsets should be provided with cut-outs to prevent sudden loud noises (acoustic shock). Headsets should also have sufficient length of cable to allow the operators to stand or move about.

Voice strain has not been intensively investigated and mostly the information is anecdotal, but it would not be surprising if there were some degree of voice strain. In many ways 'the voice' in call centres is occupying a similar role to 'the keyboard' in data entry work. A quiet workplace is one means of reducing the voice level needed, and training in speaking quietly over the telephone is also of assistance in reducing voice strain. Short breaks may help to reduce voice strain but other means must also be provided, such as an easy-to-reach supply of good-quality drinking water.

In interventions that include new equipment and work station refurbishment, the Productivity Assessment Tool can be used to assess reductions in injury costs as well as improvements in productivity.

6.4.2 *Training*

Call centres are knowledge-based industries although the technical knowledge actually required by the operators is determined by the call centres' clients, being limited to the clients' products or services, and the call centre's computer system.

There is a close relationship between the clients' requirements and training provided to operators. Training goes beyond the technical knowledge base. At the quality end of the spectrum, operators must see the requirements from the callers' points of view. They may be employed to provide a service or sell a product but, to be effective, they must give at least what the caller wants. This requires more than just knowledge of the technology of the computer system or the technology of the product; it requires an understanding of the callers and their requirements. It goes further than 'smiling down the phone', as the requirements of the callers may vary from call to call. The operator is truly responsible for the whole task.

There are three distinguishable forms of training:

- initial training when a new operator starts work
- training that occurs on a regular basis (often for one or two hours each week) to ensure that operators are aware of the latest developments, and/or when a new product or service is introduced
- individual training requirements.

Operators' requirements for initial training will vary depending on their past experience. A person new to call centres may need to learn about work station adjustment, telephone technique, the computer system, callers' requirements and selling techniques before learning about specific products or services. The longer the initial training period, the more expensive operators will be to replace.

Individual training may be for career development, or to help the operator catch up with technical knowledge of the product or service, or to enable the operator to better serve the caller – improve customer relations, the personnel side. Frequently individual training will be a task for the supervisor, and supervisors need training in the most effective way to do this; encouragement, for example, will cause less trauma in the operators than a reprimand and may produce better results.

Quality of training, in technical as well as caller (customer) relations, is clearly import-ant and poorly devised or delivered training will not produce competent operators. If operators do not feel comfortable with their level of technical knowledge or the way they handle or relate to callers, it may impinge on their health. Many call centre operators are casual employees (precarious labour) who will be under stress if they believe that, due to their lack of skill and knowledge, they are performing poorly and may lose their jobs. Even if this is an incorrect perception it does not alter the adverse health effect of a stressor. This problem is accentuated where the amount of training that precarious workers receive is reduced or limited as there is the feeling, perhaps on both sides, that the job is only temporary and short-lived.

As training is such an important facet of call centre operation and running costs, it should be entered into a cost-benefit analysis. A sensitivity analysis concerning changes to the amount of training (including amount of time supervisors spend on individual training, the number of training officers and other overheads) will identify the crucial parameters.

Labour turnover obviously has an effect on the cost of training; the higher the turnover, the higher the call centre's training costs for replacement operators.

6.4.3 Duration and rest breaks

There has been considerable research on duration of work and rest breaks in various industries. In some ways Moses was the first research worker in this field when he persuaded the Pharaohs that, if the Israelites worked six days a week rather than seven, the same amount of work would still be done in building the pyramids (our liberal interpretation of Exodus). Some time later, during the First World War, Vernon inves-tigated workers making munitions and determined that, to reduce the injury rate, there should be a break after about $2^1/_2$ hours of work (Vernon, 1921).

We have already mentioned research work (section 6.3) in which word processor operators had an increased level of musculoskeletal upper limb injuries if they worked at the keyboard for more that five hours per day. There are clearly differences between word processor and call centre operators but, just as clearly, there are similarities which implies that one should be aware of the long hours that operators may be at the telephone/computer.

Breaks during working hours have some effect in reducing the injury risks or at least reducing musculoskeletal discomfort (Henning *et al.*, 1997; Caple, 2001). To be effective, short breaks, say five to ten minutes in each hour of continuous telephone calls, have to include changes of posture by walking, stretching and so on, not just continuing sitting, and must be in addition to the longer tea/coffee and lunch breaks.

It is usual that the employer wishes to keep operators on the telephone for as long as possible; the longer the time on the telephone, the higher the productivity, or so the concept goes (see remarks on the Pharaohs above). Many call centre managers claim that the operators are on the telephones for more than 80% of the available time, but more often the figure is closer to 70% due to informal rest breaks. These 'stolen' breaks can be made formal to ensure that the operators actually rest rather than worrying that they are under observation; there would be a less tense environment and no net loss of productivity.

Changes to the rest breaks can be entered in productAbilityBasic in the 'Reduced Productivity' data entry screen.

6.4.4 Surveillance

Due to the automatic data collecting system almost all aspects of the operators' working hours can be monitored, collected, stored and remarked upon: surveillance can be the most intrusive aspect of call centre operation.

Whether and how far individual surveillance is used depends on company policy and style of management. From our observations, generally speaking there are policy differences between in-house and outsourced call centres; the latter tend to have more surveillance than the former, probably due to the commercial and competitive environment in which they operate. Too much surveillance, though, can cause a negative or 'underground' reaction from the operators leading to subversion of the employer; this is to the detriment of the health of the operators and productivity of the call centre. In many countries operators are employed outside any union involvement but, partly due to the policy of surveillance as well as other aspects of working conditions, there is a movement of operators into the union movement.

Being under surveillance is stressful for some people regardless of the purpose of the surveillance. For example, some operators have expressed resentment at surveillance with comments such as 'I have no wish that my employer should know when I have an attack of diarrhoea or am having a period . . .'. Surveillance can also be perceived to be a lack of trust by the employer of the employed. As we will see in section 6.4.5, many call centre employers go to great lengths to engender a social and cohesive feeling to the working environment but intrusive surveillance has the opposite effect, one policy negating the other.

Surveillance should *only* be used to identify training requirements of the individual operators. When used for other purposes it may lead to unnecessary stress, increased injury and absence rates, increased labour turnover and decreased productivity.

Supervisors require adequate training so as to not stress operators unnecessarily by affecting their dignity, feelings and sense of worth when discussing training requirements or information gained through surveillance.

Loss of productivity through surveillance is not easily measured and, if considered important, would need to be estimated for inclusion in the 'Reduced Productivity' data entry screen in productAbilityBasic. The estimate would need to encompass the views of the operators and supervisors.

6.4.5 Team building

Teams include a supervisor and usually 10 to 15 operators put together for administrative convenience. Many call centre enterprises believe that if a team is functioning socially it will also function productively, and the supervisor plays a major role

Figure 6.3 This is a selling team who know that work should be fun and decorate their workplace accordingly. Social cohesion ensured success for this particular team.

in developing this social cohesion. In the Deming model the supervisor has the work functions of training and support but not usually responsibility for social cohesion.

There is little doubt that teams can be effective for beginners learning from 'old hands' and in sharing information, but there can be a down-side if the social cohesion breaks down. We have already mentioned surveillance (section 6.4.4) and competition for employment (section 6.2.1) as stressors which may affect social cohesion and, in turn, lead to ill-health effects and reduced productivity.

How well a work team becomes a social team is dependent on the culture of the country as well as the personalities of the operators and supervisors; for example, the practice of holding up individuals as exemplars of good practice ('Jim has just sold another insurance policy – stand up Jim and be acknowledged and cheered by your peers') is acceptable in some cultures but not in others. For those who have sales as part of their work, if each person is judged on the sales *they* make, not the sum of the team's sales, this can reduce the social cohesion of the team. Too much importance placed on individual performance in a team can be stressful to some, although stimulating to others (see Figure 6.3).

The results of stress may be increased absence and would be entered in the 'Employee Details' data screen in productAbilityBasic.

6.4.6 Shift work and home work

Many call centres are open 24 hours a day, which may have social and ill-health consequences for operators on shift work. Working night hours, for example, interferes with the circadian (daily) rhythm and hence with sleep. Research has shown that shift work is associated with increased risk of illness, including cardiovascular disease (OHSAH, 2002). Fatigue associated with shift work has been shown to increase the risk of injury and errors at work, including risk of motor vehicle accidents.

Figure 6.4 Particularly for shift work, good kitchen and canteen facilities are essential.

As the Productivity Assessment Tool is not designed to assess the cost of long-term ill-health effects (cf. long-term diseases such as asbestosis), assessment of shift work *per se* is beyond the scope of this book. Nevertheless, one can look at the cost benefit of interventions designed to benefit shift workers in the shorter term, including, for example, kitchen and food supply facilities, night-time security and secure parking (Figure 6.4). The variables would include data entry errors, injury, absence rates and labour turnover.

An employment scheme that claimed attention some time ago, computer-based home working, has some of the elements of shift work. Although it has not been widely implemented, it was thought at the time that people would be able to work the hours that suited them best and did not interfere with, for example, family duties. The reasons for failure included:

- social; people like working with other people – working at home is isolating
- supervision; the workers are not under the direct, visual control of the employer
- training; difficulty with scheduling training at times suitable for trainers and trainees
- work station equipment is more difficult to set up away from an office and has implications for the employer's duty of care for the health and safety of the operator.

6.4.7 Software

It is unfortunate that software used in call centres is often designed for another purpose and merely adapted for call centre use. Sometimes the operators have to use several programs where, for example, speed or shortcut keys have different and conflicting meanings. This requires more training than would otherwise be needed and leads to more errors, adding to mental stress and reduced productivity.

The cost of developing new software is high, and the appeal of using what already exists is similarly high. Although the Productivity Assessment Tool is suitable to estimate the cost and benefits for developing specific software as an intervention, software design itself is beyond the scope of this book. Any analysis should consider the cost of developing software compared with the potential benefits of reduced errors, reduced stress on operators, shorter training time and improved productivity.

6.5 Case studies

Up to now we have looked at the health and safety of call centre operators, but what are the factors that describe the business side: how to make the business profitable? Sight must not be lost of this parameter but, as we have shown elsewhere in this book, there is nothing incompatible about a good, safe working environment and a profitable business.

Although the following case studies are based on actual call centres, we have used the examples principally to illustrate the financial aspects of employment. The case studies illustrate:

- employment cost parameters in call centres
- how the Productivity Assessment Tool may be used to assess employment costs.

These studies examine the financial aspects of call centre employment and are not ergonomics interventions; they are intended to form the cost basis upon which interventions may be based. As they are not interventions there are no savings or pay-back periods.

Both the case studies are from call centre enterprises that have been awarded outstanding achievement awards from their respective national call centre associations. Each case is reproduced in the productAbilityBasic program.

6.5.1 Case study 1: an outsourced call centre

The first case study concerns a company which is a typical and profitable example of an outsourced call centre placed in a large city. The major concern here is to look at the labour turnover of operators and to investigate the cost of this over and above the cost of a stable workforce.

The operators were of three categories, each category having about the same number of people. The first group of operators worked full-time, 39 hours per week; the second group worked, on average, 80% of a full week; and the third group were casual employees, working about 60% to 65% of a full week. The first two groups had all the advantages of permanent employment including paid sick and vacation leave. The third group had no paid sick leave or paid vacation leave but had an extra 12% added to their wages as recompense.

The labour turnover of the first two groups of permanent employees was low, not above 15% per year, but the third group of casual employees had a very high turnover rate, 200% per year.

There is a cost to the employer for a high turnover rate, and we have used a cost analysis to determine this additional cost for employing casual operators. The employment costs for part-time operators were, *pro rata*, the same as those for full-time operators, so that on an hourly basis the costs for part-time and full-time operators were the same.

Casual operators are usually, but not always, employed for simpler tasks and do not receive as much training as the permanent staff. Table 6.1 illustrates a typical training schedule and costs for new operators in this enterprise and shows the costs for full-time operators and casual operators. We have simplified the cost calculations by including only the direct wages, tax and social costs, direct supervision, trainers and training costs but not any other overheads. Table 6.1 is derived from the basic version of the Productivity Assessment Tool (see productAbilityBasic; File\Examples\Call centre – case 1.bgl). The full-time operators are the initial case and the casual operators are the test case. As there has been no intervention there is no pay-back period, but the Workplace Summary shows the total costs and the cost differential between the two categories of operators.

Even though not all costs were taken into account, from Table 6.1 it can be seen that the cost of employing a team of casual operators (bottom line, cost differential) is 20% greater than that of employing a team of full-time operators, due partly to the training costs.

Clearly, the reasons for a high turnover rate need to be examined. In this particular example the causes were external: the call centre enterprise has a series of short-term client contracts and the casual operators are mostly students from a nearby university wanting only short-term employment. However, if the reasons for high turnover are internal, such as stress, injury or monotony, then the enterprise could spend a considerable amount to address the problem and still be profitable.

6.5.2 Case Study 2: an in-house call centre

The following case study describes a government call centre that answers callers' questions regarding a wide variety of social security matters and requires a high degree of knowledge by the operators. In addition, this centre is rather special in that it caters for 22 foreign languages in addition to the native language. The caller tells the telephonist (not a machine) the language required and the call is directed to that language speaking person (Figure 6.5).

In this case study we are illustrating more fully the costs of employing operators. Although the call centre employed some part-time operators, making up 10% of the FTE, only the full-time operators are included in Table 6.2 and in the basic version of the Productivity Assessment Tool included with this book (see productAbilityBasic; File\Examples\Call centre – case 2.bgl).

Pay is only slightly above average for the call centre industry, but turnover is low at 15%. Most of those who do leave go to another part of the employer's workplace so that a considerable degree of promotion is available. This movement *within* an organisation is a common feature of in-house call centres, where the operators tend to associate themselves with the larger organisation rather than the call centre.

As may be expected, training new employees in this government call centre is quite intensive. Training takes place in groups of ten to twelve new employees with one trainer. Classroom training takes four to six weeks and the calculations assume an average of five weeks. Training for all operators continues at the rate of ten hours each month, and it takes about six months for a new employee to reach full competency.

Table 6.1 Costs for employing and training new operators

Details	Full-time operators[1]	Casual operators[1]	Comments
Number of operators	10	16	One team of ten FTE (full-time equivalents)
Hours of employment and wages for each operator			
Paid hours per week	39.0	24.5	
Wages paid to operators (SEK/hour)	83	93	12% extra wages for casual operators.
Paid training during employment (hours/year)	172	0	Full-time operators have 10% training during working time. Casual operators are trained only at the start of employment.
Fixed employment cost (SEK/hour)	137	159	Wages including social costs and apportioned direct supervision costs.
Training per newly employed operator			
Initial training in the training room (hours)	23	3	
Sitting alongside an experienced operator and not receiving calls (hours)	8	8	
At 50% reduced productivity (hours)	11.5	11.5	New operator sits alongside and works with an experienced operator for 23 hours. Effectively 11.5 hours lost productivity
Total time not productive (hours)	42.5	22.5	
Fixed employment cost during training per trainee operator (SEK)	$(42.5 \times 137 =)$ 5,820	$(22.5 \times 159 =)$ 3,580	Based on fixed employment costs and time during training
Annual costs including initial training costs per team of ten FTE			
Training costs for a team per year (initial employment) (SEK)[2]	$(5,820 \times 1.5 =)$ 8,730	$(3,580 \times 32 =)$ 114,560	
Employment costs for the classroom trainer (300 SEK per hour)[3]	10,350	28,800	The cost of the trainer apportioned.
Training costs for a team per year (initial employment) (SEK)	19,080	143,360	Sum of operator training costs and trainer's costs
Total costs (SEK/team/year)	2,796,000	3,390,000	**See Workplace Summary**
Cost differential (SEK/year)	–	594,000	20% increased costs for casual operators

See productAbilityBasic; File/Examples/Call centre – case 1.bgl. SEK is Swedish kroner.

1. Full-time operators are the 'Initial Case' and casual operators are the test case termed 'Casual' in the program.
2. Turnover of the full-time operators is 15% (1.5 operators per year for a team of ten operators) and for the casual operators is 200% (32 part-time operators per year for a team of 16 operators; equivalent to ten FTE).
3. The classroom trainer spends $(23 \times 1.5 =)$ 34.5 hours each year with the full-time trainee operators and $(3 \times 32 =)$ 96 hours each year with the casual trainee operators.

Figure 6.5 At the quality end of the spectrum, call centre work is varied and demands a high level of skill.

Table 6.2 illustrates the time taken to train new operators and the wage costs for the employer. In this instance the cost for the trainers has been added to the administrative costs rather than to the operators' costs; it is only a matter of how one looks at the costs – are they a cost against employing new operators or an administrative overhead? The net cost is the same.

The calculations have not taken into account the non-employment costs of the call centre, viz. head office, capital depreciation, telephone, rent and energy supplies. If these other costs need to be added they can be entered into productAbilityBasic in the Allocated Costs screen – Case costs and Equipment running costs. It is usually assumed that employment costs are 70% to 80% of gross costs.

Full details of the call centre employment costs are given in Table 6.2 so that you can use this case to determine the economic parameters that give the best financial return using the Productivity Assessment Tool.

Where would one make an intervention in this workplace? Although there is illness absence (average of 74 hours per operator per year), whether or not it is work-related is frequently not recorded. In Chapter 6.3 we referred to a Swedish study of call centres (ATK, 2001) where the research workers wished to ascertain the proportion of absence that was due to work but were unable to do so. The reason for this was the lack of detail on the medical (absence) reports sent to the employer by the physician; frequently, though, neither the operator nor the physician will realise the basic cause for an absence; for example, a cold can be ascribed to 'not work-related' but a headache or a shoulder ache could be work-related or not. There are also privacy issues between the treating physician and the patient-cum-employee that may limit the information available to the employer. The difficulty of ascribing illness absences accurately was also noted in the case study on immunisation against influenza (Chapter 5G.1).

Table 6.2 Example of costs for employing and training call centre operators

Details	Full-time operators	Comments
Number of operators	130	
EFT	130	
Details per operator		
Paid hours per week	37.0	
Illness absence (hours/year)	74	Paid illness and injury absence
Planned absence (hours/year)	215	Paid vacation and public holidays
Training during working time (hours/year)	110	This does not include training on initial employment
Productive time (hours/year)	1,525	
Wages ($/year)	38,300	
Administrative costs ($/operator/year)	6,128	Includes insurance and other costs calculated as 16% on wages
Supervisory costs ($/operator/year)	3,354	Proportion of team supervisors' wages per employee per year
Fixed employment cost ($/operator/year)	47,780	See Employee Summary
Fixed employment cost ($/operator/hour)	24.83	See Employee Summary
Training per newly employed operator		
In the training room (weeks)[1]	5	
At 50% reduced productivity (weeks)	13	New operator sits alongside and works with an experienced operator for 26 weeks. Effectively 13 weeks lost productivity
Total time not productive (weeks)	18	Effectively 666 hours lost productivity during training time[2]
Fixed employment cost during training per trainee operator ($)	(24.83 × 666 =) 16,530	Employment cost for a newly employed operator
Total costs for the call centre ($/year)		
Total employment cost ($/year)	6,211,000	Wages and other fixed employment costs for 130 operators. See Employee Summary
Training costs ($ per year)[3]	330,600	Training costs for 20 new employees ($16,530 × 20)
Recruitment administration ($/year)	18,700	
Total cost ($/year)	**6,560,000**	**See Workplace Summary**

See productAbilityBasic; File/Examples/Call centre – case 2.bgl.
1. Trainer's costs have not been included.
2. For a 37 hour working week.
3. 15% labour turnover.

To overcome these drawbacks may mean interviewing the operators one by one to ascertain whether their absence was or was not due to work and, if due to work, the reasons why. Providing this is done with the operators' understanding that the investigation is in their interests, we have not found interviewing difficult – it just takes a little time. An alternative approach, taken in the influenza immunisation case study (Chapter 5G.1), was to ascribe all short-term absences to influenza. The method used to ascertain data has to depend on the particular circumstances.

If it is estimated that half the sickness absence is due to work-related musculoskeletal discomfort, by setting up a test case in the Productivity Assessment Tool you can alter absence rates to see the effect of reduced absence on employment costs. You could include medical costs for the injured people, and also overtime worked by other operators to make up for the lost time, in addition to the wage costs for injury absence.

Although the turnover rate in this case study is quite low, 15% per year, you can calculate the savings to the government department to reduce this rate further by the similar procedure of setting up a test case in the Productivity Assessment Tool.

Yet another aspect for calculation is working hours. At present the call centre is open from 8 a.m. (08.00) to 5 p.m. (17.00) Monday to Friday but it is planned to open the call centre until 8 p.m. (20.00) each night, Monday to Friday. What is the cost benefit, or even cost loss, of employing extra operators to work these evening hours rather than the existing operators working overtime? If there already is a degree of absence due to work-related causes (say musculoskeletal discomfort of the shoulders and neck), it would not be in the operators' best health interests to work longer hours even if they wish to work overtime. It is well recognised that injury causation, particularly musculoskeletal injury, is related to duration of work.

The calculations can be made using the basic version of the Productivity Assessment Tool, although a sensitivity analysis is more easily made using the complete version where you can add part-time and full-time employees as well as comparing alternative interventions for injury prevention at the same time.

Of course, calculating the financial parameters has nothing to say about which intervention would be appropriate to address health issues, but it can assist with determining the cost-effectiveness of the intervention.

6.5.3 Conclusions

Any intervention needs to be aimed not only at improving working conditions to prevent injury but also at the cost-sensitive parameters. These cost parameters depend on, for example, injury absence, labour turnover, training costs and productive time. These facets are also linked with the degree of complexity of the call centre operators' tasks; the more complex the work, the more training is required and the greater the negative cost effect of high labour turnover. Although these are self-evident factors, they are the parameters that need to be addressed to make an intervention financially successful as well as successful in injury prevention terms.

The case studies in this section are reproduced in the basic version of the Productivity Assessment Tool included with this book, and they can be used to illustrate the various employment parameters when costing call centre operations or any other computer-based work.

6.6 References

ATK, 2001. *How to Develop Work in Call Centres – a Summary of the Results and Conclusion from the Project 'Call Centres in Development – Long Term Sustainable Work with Customers at a Distance.* Obtainable from ATK, Box 17508, 11891 Stockholm, Sweden.

Bain, P. and Taylor, P., 2000. Entrapped by the 'electronic panopticon'? Worker resistance in the call centre. *New Technology, Work and Employment*, 15, no. 1, 2–18.

Caple, D., 2001. Ergonomic review of rest breaks in call centres, in Stevenson, M. and Talbot, J. (eds), *Proceedings of the 37th Annual Conference of the Ergonomics Society of Australia* (Canberra: Ergonomics Society of Australia), pp. 91–96.

Health & Safety Executive, 2001. *Local Authority Circular.* See http://www.hse.gov.uk and go to 'call centres'.

Henning, R.A., Jaques, P., Kissel, G.V., Sullivan, A.B. and Alteras-Webb, S.M., 1997. Frequent short rest breaks from computer work: effects on productivity and well-being at two field sites. *Ergonomics*, 40, no. 1, 78–91.

Ishikawa, I., 1985. *What is Total Quality Control?* translated by Lu, D.J. (Englewood Cliffs, NJ: Prentice Hall), p. 59.

Norman, K., Toomingas, A., Nilsson T., Hagberg, M. and Wigaeus Tornqvist, E., 2001. 'Working conditions and health among female and male employees at a call center in Sweden'. *Fourth International Scientific Conference on Prevention of Work-Related Musculoskeletal Disorders (Premus)*, 30 September–4 October, Amsterdam.

OHSAH, 2002. *Shift Work – a Review of the Literature*, Occupational Health & Safety Agency for Health Care in British Columbia, http://www.ohsah.bc.ca/chemical-exposure-monitoring.html#injury.

Oxenburgh, M.S., 1984. Musculoskeletal injuries occurring in word processor operators. *Proceedings of the Ergonomics Society of Australia and New Zealand* (Sydney), pp. 137–143. Reproduced in Stevenson, M. (ed.), 1987. *Readings in RSI. The Ergonomics Approach to Repetition Strain Injuries*, (Sydney: New South Wales University Press), pp. 91–95.

Taylor, P. and Bain, P., 1999. 'An assembly line in the head': work and employee relations in the call centre, *Industrial Relations Journal*, 30, no. 2, 101–117.

Vernon, H.M., 1921. *Industrial Fatigue and Efficiency* (London: George Routledge & Sons).

Winkle, J. and Oxenburgh, M., 1990. Towards optimizing physical activity in VDT/office work, in Sauter, S., Dainoff, M. and Smith, M. (eds), *Promoting Health and Productivity in the Computerized Office* (London: Taylor & Francis), pp. 94–117.

7 Ethics

Contents

7.1 Why ethics?

Ethics plays a role in all day-to-day interactions between people, even though we are often not aware that it does so. In our dealings with our children, friends and relatives we take it for granted that we should be fair and honest. These ethical issues are extended into our working lives, for, if we are honest at home, we do not expect to be dishonest at work. However, there is sometimes a clash between ethics and culture, and that is why we feel that the issue needs to be addressed here.

Interventions in a workplace usually address culture, and may even attempt to change a culture, but are the means to do so ethically correct? This chapter examines the question of the ethics of ergonomics intervention in the workplace and, as this book is also concerned with finance, the role of cost-benefit analysis in the intervention – whose costs and whose benefits?

To use the *Macquarie Dictionary* definition, ethics is 'a system of moral principles, by which human actions and proposals may be judged good or bad or right or wrong'.

However, the second definition is 'the rules of conduct recognised in respect of a particular class of human actions'. In this chapter it is intended firstly to look at some of the principles by which we intervene in the workplace; that is, the moral principles of our conduct. In the second part the principles will be used to see if our rules of conduct (for example the professional societies' codes of practice) conform with the principles.

For convenience the terms *ergonomics* and *ergonomists* will be used, but the same arguments apply to other areas of workplace health and safety, to other practitioners and those concerned with workplace interventions.

It should also be pointed out that these are the reflections and cogitations of the first author, Maurice Oxenburgh. Some of the arguments discussed here were presented to an annual meeting of the Ergonomics Society of Australia (Oxenburgh, 2002).

7.2 Cost-benefit analysis

When I wrote the first edition of this book Professor Tom Armstrong, from the Center for Ergonomics at the University of Michigan, questioned the basis of putting cost-benefit and workers' health together. His argument was that we should expect a safe and healthy workplace independently of cost for, if we use cost in the discussion, there is the counter-argument that 'if it is not financially successful, why do it?'. This question worried me and remained at the back of my mind, and I now feel that I should address this issue.

Before discussing the question of using a cost-benefit argument to support a workplace intervention, it will be as well to examine some of the assumptions built into a cost-benefit model. The general economic assumptions are described in Chapter 2.5.2 and include, for example, 'there is "perfect" knowledge for consumers, sellers and workers'. This is as patently untrue as the assumption that follows, 'workers make fully informed decisions about their employment conditions'.

It is taken for granted in discussions about ethics that *truth* is the overriding factor and must not be subordinated. If this is the case, how can we use a financial argument (a cost-benefit model) based on the above two untrue assumptions, let alone the several that follow in Chapter 2? Knowledge is neither equal nor complete and workers are rarely informed about the hazards of their employment. Of course, these are only assumptions made to simplify an otherwise complex system; the system as a whole cannot be modelled on the entire truth but only on a simplified formula. Given that this is so, do simplifications and assumptions make a model valueless or even ethically bankrupt?

Providing we know that the model is a simplification for convenience only and is not designed to detract from the truth, then the model is useful: an argument implied throughout this book. The question is not whether or not the cost-benefit model is useful in its stated objective of convincing management to implement ergonomics interventions; the question is whether the simplifications and assumptions make the concept of the model ethically bankrupt. That a concept is *useful* is not the same as it being *ethical*. Or, to put it the other way round, is it ethically justified to use arguments based on these false or simplified assumptions?

One key question is our use of cost-benefit analysis. In my own mind I have overcome this problem by saying that it is only one of several methods to stimulate management to improve workplace conditions. In the past, ergonomists and health and safety practitioners generally had at their command only limited methods for such encouragement:

- legislation, but that is limited in scope and does not assist where there is not an immediate or obvious danger of injury
- workers' compensation premiums, which are often taken as being part of the normal running costs or, in some countries, are fixed and are no stimulus to a better or safer workplace
- industrial relations, where health and safety issues are often subordinated by, or subordinate to, other issues such as wages, working hours and power relationships
- enterprise image, which can be an important factor particularly for well-known companies that rely, at least partly, on advertising and public relations for their income but is of limited utility for workplace ergonomics
- an appeal to the humanity of the managers, which sometimes works but is often diminished when it comes to allocating resources.

Thus, financial analysis is just another tool.

However, if legislation is accepted in the culture of a community for regulating safety in the workplace, is that the same as legislation being ethically acceptable? If the legislation is coincident with ethical principles, and if cost-benefit analysis is just another tool to 'support' legislation to implement ergonomics, then there is no ethical issue. However, the consequences of some legislation, especially that relating to industrial relations and employment, may not be ethically justified if they adversely affect some people in the community. If our cost-benefit argument is simply and only an alternative argument to unethical legislation, then our argument in favour of cost-benefit analysis being ethically acceptable fails, at least in this instance.

7.3 Precarious workers

There is a continuing rise in the number and proportion of precarious workers in the workforce of all countries and, if ergonomics interventions are assisting in this by increasing productivity, are we but the hand-maidens of the worst aspects of capitalism?

As an aside, I do not subscribe to the argument that an increase in the proportion of precarious workers is good for the economy on the basis that if frees up restrictive labour practices. I tend to believe that the individual good is also important and one needs to ask the precarious workers for their views!

If you use the argument that it is cost-effective to implement certain ergonomics improvements there are going to be cases whereby the employer can get the work done with fewer workers. But you can also say that there is, for example, a legislative requirement to implement certain safety improvements and this may lead to the same outcome; fewer workers needed. That is, cost-benefit analysis is only a tool and we could have used this other argument (e.g. legislative requirements); it is not necessarily the *argument* we use to assist in the implementation of an ergonomics intervention, it is the *consequences* of the ergonomics intervention that is the ethical issue.

Assuming for the sake of this discussion that we feel that the use of cost-benefit analysis is not unethical, or at least ethically neutral, if used for its limited objective of encouraging management to implement an ergonomics intervention, that still leaves the most important questions unanswered: the implementation of ergonomics and the consequences that follow.

7.4 Implementation of ergonomics

I feel that Professor Armstrong's question has addressed only part of the ethical problem. In my view the question should not be is it ethically correct to use financial analysis to implement change, but what are the *consequences* of change? I would not go so far as to claim that the means justify the end, but within the limitations and prerogatives of a capitalist culture financial analysis is a sensible reason for action.

If we are to look at the consequences rather than the reason for an intervention (the motivation of management), the question becomes: is ergonomics in the workplace there to prevent injury and increase the well-being of workers, or is it there to increase the profitability of the enterprise? To my way of thinking, the overriding issue is that of the consequences of the intervention and not the rationale (whether cost-benefit, legislation, etc.) for implementation.

Chapter 5A illustrates a case study whereby efficiency of industrial cleaning was markedly increased and the time to do certain tasks reduced. This meant that fewer people were needed and the employer, if it had so chosen, could have reduced the number of people employed. Yes, there was a reduction in injury potential but there was potentially an increase in unemployment, a situation that is far from uncommon.

As ergonomists we become part of the problem and not always the solution, as we usually think we are. Is an ergonomics intervention in the workplace always a good thing if it reduces injury yet sometimes increases unemployment? Before examining this issue it may be as well to think about the issues that an ergonomist will come across when asked to look at a workplace and the potential conflict of interests.

7.5 Conflict of interests

Almost inevitably when we are asked to assess a situation we are faced with the question, who am I working for? Is it my boss or whosoever will pay for my work and/or the person who has asked me to assess the workplace, or the people who work at the workplace – the workers?

It is quite clear from the codes of practice (dealt with later in this chapter) that we are responsible to the community and to the workers. But is this always, or can this always be, true? We need to earn a living. If employed, we need to do at least some of our employer's bidding; if self-employed, we need to ensure continued business. In practice we are usually responsible to the employer who pays us, rather than to the employed who is at risk. 'He who pays the piper calls the tune.'

Thus, in a roundabout way, we may be using the argument that we wish to increase the safety of workers while actually supporting our own employment as ergonomists. If this is the case (and I hope it is not) then there may be an ethical divide between the ergonomist's intent and the well-being of the workers. Hopefully the two will coincide, but not always.

But all is not gloom and despair. In any business transaction there is a degree of choice. If we look at the ergonomist as a business person doing business with another business person (the employer), then the transaction needs to satisfy both parties, preferably both gaining equally, but the gain does not always have to be financial. When implementing an ergonomics solution in a workplace, my satisfaction does not depend solely on whether they will pay me for the work, but also on whether I enjoyed the task, whether I did it properly and whether I achieved a satisfactory result for the workers at risk of injury. Similarly, the other business person, the employer, will be satisfied if there is a gain, and that may not always be financial; it could be that he or she has a better relationship with the employees. Thus, although I have to ensure continuing business, it is not the only objective. This is discussed interestingly and in more detail by Griffiths and Lucas (1996).

Nevertheless, conflict of interests is an issue that needs to be sorted out at the beginning of a project, not at the end.

7.6 Intent

We come to the issue of intent, and I believe that this is the nub of our argument. To sort out a potential conflict of interests must lead us to the intent of the intervention. What did we intend the consequences to be when we were asked to make an ergonomics intervention?

If we look at Chapter 5B, the hospitality industry case study, there can be no doubt that the initial intention was to reduce injury rates. That the improvements allowed the employer to reduce full-time employment and increase casual/precarious employment was not foreseen at the beginning of the project by either management or the ergonomists involved. There seems to be an unstated assumption that intentions are always honourable but the outcome of an intervention may not always be desirable, at least for the employees.

If the intent is clearly stated, understood and agreed to at the beginning of an intervention then rationally the outcome, even if undesirable, cannot detract from the original intent.

7.7 Responsibility

However, I have some disquiet that, despite my good intentions, a few of the ergonomics interventions that I have been concerned with had consequences which I did not intend; due to improved working conditions there was reduced injury leading to increased productivity *but* also leading to a smaller workforce. Where does responsibility end? We cannot look into the future, so does our practical responsibility end when we have finished the ergonomics intervention? Or even when we have stated our intent at the beginning?

If we accept that our responsibility is limited in time, which seems rational, and our intent has been stated at the outset, then the responsibility for our intervention leading to unforeseen or unwanted consequences does not arise. It is only if an intervention had the foreseeable consequences of, for example, unemployment that there is an ethical issue, although in some cases our responsibility may be modified by other factors related to the ease of displaced employees obtaining other employment.

But what about the responsibility of the employers? Some time ago the first two authors were asked by the workplace inspectorate in New South Wales to prepare a booklet of case studies to illustrate the cost-benefit approach (WorkCover, 1999). The initiative for this did not come from my ergonomics or safety colleagues but from industry itself. There was a view that, 'yes we would like to improve our workplaces and make them safe, but how can we afford to do it?'.

In our market/capitalist economy we are used to the profit motive reigning supreme, but here is industry not saying 'how can we make more money?' but saying 'we are concerned with the humane approach, so please tell us how we can afford it'.

One of the best managers I have come across was the manager of a heavy industry factory, and his 'nightmare' was having to go to an employee's home and tell his wife that she was now a widow. He did not need any financial incentive, his humanity and imagination were sufficient, and that is what kept the factory in good and safe shape; one could literally eat off the floor!

Managers as individuals in direct contact with workers do accept responsibility, but it is frequently tempered by perceived financial restraints where ethics is competing with the workplace culture of profit and cost-cutting.

In large corporations management and boardroom decisions may be taken at a distance from those affected (arm's length or remote decision-taking), which makes it easier for managers to ignore (or not even know) the consequences of their actions. However, this does not dilute their ethical responsibility.

7.8 The means or the end?

Slowly I am weaving a web around the problem that Professor Armstrong set and coming to the conclusion that the ethical question does not concern cost-benefit analysis but the actual ergonomics intervention. He was concerned that the *means* of implementation, the financial arguments, becomes the *end*, which is profit as far as an enterprise is concerned.

To my way of thinking, the ethical issues are not those of using financial methods to encourage management to implement ergonomics but the possible consequences arising from the implementation of ergonomics. We need to look at *conflict of interests*, *intent* and *responsibility* of those implementing, or advising on the implementation of, the changes (see Figure 7.1).

Even if the *intent* of an ergonomics intervention was 'good', at least in terms of the intended improvements to the health and well-being of workers, it may well have negative consequences, for example people unemployed. Unless we know that unemployment will be an outcome we will have discharged our *responsibility* by our suggested ergonomics intervention. If unemployment or other negative effects is a foreseeable outcome, then we must be honest about that, but to whom; the employer who pays us or the workers who may be affected but with whom we have no business or employment contract? This is a major reason for sorting out any conflict of interests before the project begins rather than after.

I know that I have not addressed all the potential issues and the ones I have addressed have only scratched the surface, but we have defined at least some of the places where we need to implement ethical principles. We, or at least I, can certainly claim that to use cost-benefit analysis to implement an ergonomics intervention can be justified if we have sorted out the difficulties of conflict of interests, intent and responsibility. On that assumption we can now turn to what principles we actually need to ensure an ethical basis for our work.

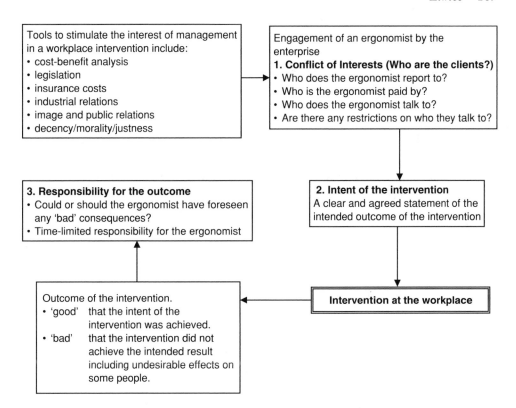

Figure 7.1 The three areas where ethical issues may arise for an ergonomist (or an occupational health and safety practitioner) before and after a workplace intervention.

7.9 Principles of ethics

Most professional societies produce Codes of Practice, sometimes called Codes of Ethics. However, we should be quite clear about the difference between ethics and codes of practice. The former is the basis upon which the latter should be based; each item in a code of practice should be consistent with the ethical principles on which it is based.

Although ergonomics is rarely a matter of life and death, that certainly is the case for some medical practitioners. The Appleton Consensus (Stanley, 1989) was a meeting of over 30 physicians, bioethicists and economists from ten countries who met to draw up ethical principles upon which to base certain actions; that is, when to 'pull the plug' on patients who are not competent (unconscious) and who will never be competent but are being kept alive by medical technology. I have used some of their ethical principles and adapted them for ergonomists, but I make no claim that these are complete or even the best ones to choose from. I will briefly discuss five ethical principles and then relate these to the codes of practice. The first four principles are adapted from the Appleton Consensus and may be summarised as follows:

- **do no harm**; a moral obligation not to harm others
- **do good**; a moral obligation to assist those who need assistance

- **equity**; fairness and impartiality
- **justice**; to act fairly and with truth.

How do these relate to ergonomists and occupational health and safety practitioners in general?

7.9.1 Do no harm/do good

That is an easy one for ergonomists. We always do good and never any harm, or do we?

To take our example in Chapter 5A concerning industrial cleaners; there had been a reduced risk of injury and productivity had been increased (a good job was done) but now only three people are needed, leaving one person with no job (harm has been done). In this actual case study the 'redundant' cleaner was employed elsewhere, but this is not always the case and could not have been anticipated ahead of the actual occurrence. In terms of our previous argument, the intent of the intervention was clearly defined and had been met but there were undesirable consequences that were not foreseen.

7.9.2 Equity

This, along with justice, leads to the conflict of interest: **who is the client?** With a physician the client is the patient with the broken leg, but who is the ergonomist's client: the person at risk of injury (the worker) or the one who pays the bill (the employer)? To be fair to one may be unfair to the other. To use a hackneyed expression, if there is a 'win–win' situation where each side profits equally, there is no unfairness. But if one side perceives that there has been a loss, or an unequal gain by the other, there is no equity; there is no fairness.

Paul Branton, an English ergonomist, (Oborne *et al.*, 1993) conceived the concept of person-centred ergonomics. He started with the concept of the whole person – the individual's psychology, physiology and philosophy. Branton saw clearly that the client is the person or persons at risk and most of us would intuitively agree with this. However, most of us also believe that ergonomics is concerned with the whole system, the man–machine system. Branton went one step further to consider the person first. Should we go for the whole system, the man–machine system, or for the person-first system?

The codes usually put the person (worker) first, but the practice of much of ergonomics is concerned with the whole system. There is this potential conflict in the International Commission on Occupational Health code, where it is stated that the practitioners should 'contribute to environmental and community health'. The economic health of a community is surely important, but this can be at variance with the interests of some individuals within the community: it is clearly in the interests of a company and its employees (together making one 'community') to survive in a tough economic climate but, by an increase in productivity due to an ergonomics intervention, this may mean fewer people employed in other communities. In this event only one side 'wins' there is no equity for the others.

There is perhaps an exception to this concept of equity being fairness to all involved: the exception is to skew the benefits towards the 'underprivileged'. This is sometimes a policy of enlightened governments towards minority or discriminated-against groups and is a form of positive discrimination. One can make a case that workers are an underprivileged group and that the aim of health practitioners should be directed towards positive discrimination in their favour. The codes of practice

---

Basic principles of the International Commission on Occupational Health Code of Ethics

The following list summarises the principles of ethics on which is based the International Code of Ethics for Occupational Health Professionals prepared by the International Commission on Occupational Health (ICOH) (ILO, 1998).

- Occupational health practice must be performed according to the highest professional standards and ethical principles. Occupational health professionals must serve the health and social well-being of the workers, individually and collectively. They must also contribute to environmental and community health.
- The obligations of occupational health professionals include protecting the life and the health of the worker, respecting human dignity and promoting the highest ethical principles in occupational health policies and programs. Integrity in professional conduct, impartiality and the protection of the confidentiality of health data and of the privacy of workers are part of these obligations.
- Occupational health professionals are experts who must enjoy full professional independence in the execution of their functions. They must acquire and maintain the competence necessary for their duties and require conditions which allow them to carry out their tasks according to good practice and professional ethics.

---

seem to assume, without specifically stating, this to be the case but the concept needs careful assessment rather than a blanket assumption.

7.9.3 *Justice*

Three people or parties are involved in an ergonomics intervention – ergonomist, employer, employee. Can all three be treated justly? There may be differences not only in education, outlook and class structure between individuals but also in their relative positions (the power structure) within an enterprise. If there is no equity between employer and employed how can we expect there to be justice?

We, as ergonomists, may overcome some of this by engaging in worker participation but, due to unequal knowledge, power and position, it is not necessarily successful. Preliminary research work has indicated that some aspects of worker participation actually lead to stress in some workers; if a person has enough problems from work and/or private life then the extra task of worker participation may be too much.

Worker participation also has the pitfall that ergonomists may become part of a management game or be seen to be part of management rather than recognised as independent advisers.

Where, then, is the equity and justice between the worker/citizen and the community/employer, and can there be equity and justice? Ergonomists may become circumscribed by community (workplace) economics, which is quite different from using financial arguments to support an ergonomics intervention. If an ergonomist is engaged on a project designed to support the enterprise's finances (*intent*), there is the danger that one becomes biased to one side, the employing side, and is acting outside the role of the independent specialist.

7.9.4 Self-happiness

Bertrand Russell (Pigden, 1999) believed that ethics is active and self-centred. One's actions must satisfy one's inner soul or feelings, although clearly limited by other principles: pride, not arrogance. Putting this into our context, unless we are happy being ergonomists we may not do a competent job. I am not sure that I agree that self-happiness is an ethical principle, but it is at least a concept that is essential to good professional practice. I can find nothing in the codes of practice that states that I should enjoy my work!

So here are some ethical principles. They are not complete and, moreover, in any situation there may be conflict between their interpretation and the weight given to any particular principle. But we must start somewhere and, for the purposes of looking at codes of practice, perhaps they make a reasonable starting point.

7.10 Guidelines and codes of practice

If we can agree that the ethical principles upon which we should base our workplace interventions are those of conflict of interests, intent and responsibility, then how do we fare? Are the codes of practice from the various societies sufficient and do they relate to the ethical principles above, or even to other ethical principles?

The general guidelines prepared by the International Commission on Occupational Health state its basic principles, which are the basis for the guidelines (see box). The guidelines also have a section entitled 'Conditions of execution of the functions of occupational health professionals' where it is stated that 'Occupational health professionals must always act, as a matter of priority, in the interest of the health and safety of the workers'.

With the physicians the issue of conflict of interests is relatively clear-cut. There is the tradition of responsibility to the patient with confidentiality between the patient and the physician accepted; frequently there is legal protection for the physician and this makes it is easier for the occupational physician to be responsible to the patient-cum-worker without a conflict of interests with the community. This tradition does not always encompass ergonomists. If we define the enterprise as the community, ergonomists are left with no clear guidelines on whom we are responsible to, the worker or the employer, when there is a conflict of interests.

If we have decided that the crucial aspects for ergonomists are conflict of interests, intent and responsibility, what guidelines do we have from the professional ergonomics societies? In general the guidelines do tackle the issue of responsibility and come down on the side of responsibility to the community and the worker (which are not always identical), but there seem to be some exceptions where responsibility is towards the employer.

Rather than discussing all the codes, the Ergonomics Society of Australia's Code of Practice will serve simply as an example. It is a short code, a definite advantage, and others do not necessarily go any further in addressing ethical issues.

Code of practice – an example

As a condition of admission to professional grades, members of the Ergonomics Society of Australia Inc. agree to the following statement (ESA, 2002).

When practising ergonomics, members shall, at all times:

1 ensure that the community and clients' well-being take precedence over their responsibility to sectional or private interests
2 uphold and enhance the honour, integrity and dignity of the profession and of the members of the Society
3 ensure that their responsibility for the ethical conduct of any behaviour involving representation of the Society has due regard for the professional integrity of the Society; that this responsibility is not used to abuse the privilege; and this responsibility takes precedence over any concern for the sectional, private or commercial interests or advantage
4 express opinions on the work or reputation of fellow members in an honest, objective and responsible manner, giving due credit where necessary
5 provide advice, express opinions or make statements honestly, objectively, impartially, expeditiously and reporting on the positive and the negative consequences of that advice
6 perform work only in their areas of competence and to the best of their ability
7 disclose to their employers or clients promptly and effectively all significant financial and other relevant interests with potential for providing conflict of interest or influencing the impartiality of any reports, advice or decisions
8 respect the confidentiality of the information obtained in the course of their work as ergonomists, revealing such information to others only with the consent of the person(s) or organisation(s) or their legal representative(s)
9 actively assist and encourage the ongoing development of ergonomics
10 agree that non-compliance with the Code may be referred to the ESA Board for determination.

There is no preamble to explain the ethical basis of the code, which is also the case with the Ergonomics Society (ES, UK) and the Human Factors and Ergonomics Society (HFES, USA) codes. It is expected that community obligations and honesty are the first considerations and that there should be no conflict of interests; all well and good. The HFES for convenience separates the various roles of ergonomists as academics publishing articles, as expert witnesses in legal proceedings and so on, but does not explain whether there are different ethical issues in the various roles and there is no reference to any ethical principles.

I think that the codes do attempt to address responsibility by putting the community and worker first but assume that these interests will always coincide; that there will be no conflict of interests between the community, however defined, and the worker. There is no attempt, or at least no clear attempt, to state clearly what the intent (in ethical terms) of an intervention should be.

That these may only occasionally be issues is not the point. I am sure that all of us who have been working in occupational health and safety for some time have come across ethical conflicts in ourselves and/or in others. I have certainly come across this conflict in my work and, in retrospect, on at least one occasion went the wrong way, and I am not at all sure that the codes of practice would have 'saved' me. The conflict was one of 'who was my client?' and reference to ethical principles might have made the issues clearer.

The preamble of the HFES Code actually states that: 'No such code can be expected to completely anticipate all of the various and complex arrangements that professionals create, nor can it fully explore the many ramifications of these arrangements', which, I think, supports my contention that the codes must be based on ethical principles so that:

- firstly, the code of practice has a stated ethical basis
- secondly, unanticipated situations can be addressed by reference to the ethical principles.

There is an important third reason for having a set of ethical principles: it avoids the error of going according to a set of guidelines that may become outdated but not updated; the principle must always be more important than the word.

7.11 Conclusion

So what is the conclusion? As in all ethical and philosophical debates, there is no conclusion, only a continuing discussion. In this light I have only raised some of the issues and it is up to each of us to sort out our issues of conflict of interests, intent and responsibility. The various codes of practice are useful but do not necessarily address these ethical issues, as I believe they should.

In my view the concept of cost-benefit analysis, of itself, is not an ethical issue provided that one is firstly cognisant of the ethical issue of intent, as well as of a conflict of interests. Cost-benefit is a tool and has no ethical value; the manner in which it is used is the ethical issue.

Thus I have confined myself to discussing the approach that an ergonomist (or other occupational health and safety practitioner) should make before and after an intervention. There are other ethical issues which should be addressed:

- the relationship between practitioners when competing for work (we all need to eat)
- acting in a legal environment, for example as expert witnesses in court
- responsibility to the community (general public) on occupational health issues
- better and safer design of equipment and buildings.

I will leave these for another day.

7.12 References

ESA, 2002. *Ergonomics Society of Australia Directory* (Canberra).
Griffiths, M.R. and Lucas, J.R., 1996. *Ethical Economics* (London: Macmillan).

ILO, 1998. *Encyclopedia of Occupational Health and Safety*, 4th edn. (Geneva: International London Office).

Oborne, D.J., Branton, R., Leal, F., Shipley, P. and Stewart, T., 1993. *Person-Centred Ergonomics: a Brantonian View of Human Factors* (London: Taylor & Francis).

Oxenburgh, M.S., 1991. *Productivity and Profit through Health & Safety* (Sydney: CCH).

Oxenburgh, M.S., 2002. Ergonomics, economics and ethics, *Ergonomics Australia*, **16**, 12–14.

Pigden, C.R., 1999. *Russell on Ethics. Selections from the Writings of Bertrand Russell* (London: Routledge), pp. 151–163.

Stanley, J.M., 1989. The Appleton Consensus: suggested international guidelines of decisions to forgo medical treatment. *Journal of Medical Ethics*, **15**, 129–136.

WorkCover, 1999. *Linking productivity and OHS: a quick guide to costs and benefits.* [Copies may be obtained from WorkCover, GPO Box 5368, Sydney, NSW, 2001, Australia and quoting catalogue number 753.]

8 Installing and running the Productivity Assessment Tool

The software for the Productivity Assessment Tool is known as **productAbilityBasic**. This chapter gives the technical details for the installation and operation of the software.

System requirements

productAbilityBasic requires Windows™ NT/95/98/2000/ME/XP.

If you have Java already installed you will require 3MB of disk space. To install the Java VM (virtual machine) you will require about 40MB of free disk space.

Installation of productAbilityBasic

Insert the productAbilityBasic CD in your disk drive and the installation program should start immediately. If it does not then open an Explorer window, go to your CD-ROM drive, click on install.htm and follow the instructions from there.

Installing Java™ separately

To run productAbilityBasic you require the VM to be installed. To reduce the size of the install you may wish to use a previously installed version of Java™. If there is one on your computer and it is the correct version (Java 1.4.1) then no VM is required. If no VM is available you will have to either install a VM or use the 'install VM' option on the supplied CD. This will install a dedicated VM that will be available only to productAbilityBasic. However, if you move, remove or upgrade Java™ you may have to reinstall productAbilityBasic.

Starting productAbilityBasic

To run the productAbilityBasic program, open the folder \Program Files\ productAbilityBasic and click on the productAbilityBasic shortcut icon.

Ordering the complete version of the Productivity Assessment Tool

Accompanying this book is the software for the *basic* version of the Productivity Assessment Tool (see Chapter 4). This basic version is restricted to the initial case and one test case and only one employee group; the *complete* program allows up to four interventions (test cases) and up to five employee groups within the same intervention. The complete version thus allows considerably more flexibility and it may be ordered by request to the authors.

For more details please see www.productAbility.co.uk or send an e-mail to maurice_oxenburgh@compuserve.com.

Glossary

Install VM	An option allowing the user to install a VM especially for them. The VM will be placed in the same directory as productAbilityBasic and will be removed if uninstalled.
Java™	An operating environment designed and maintained by Sun Microsystems, Inc. It is a machine-independent system and allows programs built in one operating system to be run seamlessly on another operating system.
VM/Virtual Machine	The use of Java™ requires a local Virtual Machine (VM) which is specific to the Operating System. Java is a registered trademark of Sun Microsystems, Inc.
Uninstall	If you need to remove productAbilityBasic there is an uninstall program. This will delete all installed files. You will still have to delete the directory as there are some files that may not be removed automatically.

Glossary

Definitions specific to the Productivity Assessment Tool will be found in that program (see Help/Definitions).

accounting system	Refers to enterprise-based systems to account for wages and other expenses. 'Financial system' is an alternative term.
assumptions	The principles, sometimes unstated, upon which a theory is built. In economics theory they are estimates, guesses or generalisations about how groups of workers, industries or economies will behave.
bottom line	The last line of a financial statement of a company; profit and loss. May be used in a derogatory sense to indicate a limited vision of the senior enterprise management.
call centre	Uses computer technology in tandem with telephone technology to form a new system. It is essentially the technical integration of a communication system (the telephone) and a rapid retrieval data base system (the computer). Originally developed to answer callers' questions but now also developed to sell goods by 'cold calling'.
call centre operator	One who works in a call centre on the telephones.
caller	An outside person who telephones into the call centre or who is telephoned by the call centre.
casual employee	One who is employed on a non-permanent basis. *See* **precarious work/workers**.
cervicobrachial disorder	A musculoskeletal disorder of the shoulder and/or upper limbs. *See* **musculoskeletal injury**.
ceteris paribus	'All else remaining the same.' Used in economic models to indicate that the model is based on the standard economic assumptions.
client	1. The enterprise which is employing an outsourced call centre for services.
	2. The person(s) who employs the ergonomist and/or to whom the ergonomist may have responsibilities. The term 'stakeholder' is a wider term and may not be applicable in the limited sense of a workplace intervention.
cold calling	From a call centre; telephoning outside persons but not necessarily at their request.

confounding factors	In an intervention, factors that confuse, interfere with, or dominate the factor being measured.
cost-benefit	Any situation where the costs and benefits of an intervention are compared, as distinct from cost-benefit analysis. *See also* **cost-effectiveness, net benefit.**
cost-benefit analysis	Refers to a model that purports to identify all the costs and benefits associated with the proposal under consideration, including society-wide costs and benefits. This is an assumption in itself, as it is not feasible for all costs and benefits to be known. Estimates of the known costs and benefits are quantified in monetary terms and then directly compared with each other.
cost-effectiveness	Is used in *any* situation where the costs and benefits of a proposal or intervention are compared and where the benefits exceed the costs, as distinct from cost-effectiveness analysis. *See also* **cost-benefit, net benefit.**
cost-effectiveness analysis	Compares the costs and benefits of a proposal or intervention to the enterprise, including the social and cultural, and is a macroeconomic model. This type of analysis is useful where factors external to individual enterprises need to be included or where costs or benefits cannot reasonably be quantified.
CTD	*See* **cumulative trauma disorder.**
cumulative trauma disorder	A work-related injury, commonly due to overuse, usually of the upper limbs. *See also* **musculoskeletal injury.**
dB(A)	A decibel scale with the A frequency weighting used in the measurement of sound pressure levels (noise).
demand	Equates to buyers of a product or service. In economic theory it is the quantity of goods or services that will be sold at any given price.
Deming	In the Deming model workers work as a team and the team members pool their ideas; the team as a whole is acknowledged for its efforts and ideas. From W.E. Deming, author of many books on production methodology in industry including *Out of the Crisis* (Cambridge University Press, 1982).
direct costs	Gross wage costs, including provision for annual leave, sick leave, superannuation/pension, social and other taxes.
discount rate	Costs may be increased to account for inflation and interest rates and benefits may be discounted by an amount to take into account the need to wait to enjoy them.
economic model	Used to test economic theory to particular situations, usually to predict what will happen in the future. The models represent the economic world, or a part of that world, but are simplified derivatives. They are mathematically based and work by translating economic factors into numerical values so they can be added, subtracted, multiplied or divided together to give a numerical

	answer. Used as a general term to include accounting and financial models. *See also* **accounting system.**
economics	The science of the production, distribution and use of income and wealth.
employee	A person employed for payment. *See* **labour.**
enterprise	A company, organisation, government department, etc.
externalities	Situations where some of the costs of producing goods or services are external to, but not paid for, by the enterprise. For example, pumping waste into local rivers and not paying to clean the water or dismissing injured workers to avoid paying compensation and other costs. These costs are then borne by the state or others.
financial appraisal	A method of assessing a proposal in financial terms. It considers the cost of a proposal, often termed an outlay or investment, and then calculates the benefits from the proposal taking into account the time frame which it will take to recover the initial outlay or investment. *See also* **return on investment.**
financial system	*See* **accounting system.**
FTE	*See* **full-time equivalent employee.**
full-time equivalent employee	The sum of the hours of people working in an enterprise, or part of an enterprise, divided by the hours worked by a full-time employee. For example, two people working 20 hours each per week will be the equivalent of one full-time employee working a 40-hour week.
Hawthorne effect	The name given to a change in the working conditions or working environment that has an initial effect but does not last after the newness or novelty has worn off. The name is derived from a workplace experiment at the Hawthorne Works of the Western Electric Company in the 1920s.
hazard	A situation where there is the potential for an injury.
hidden cost	The indirect costs that an enterprise may have built into its cost structure. Often, these costs are not separately identified and are termed 'hidden'. *See* **indirect costs.**
hot-desk(ing)	The same desk used by several people although usually on different shifts.
indirect costs	Costs that are part of the production of goods or services but exclude the direct costs of wages, workers' compensation premium, social and other taxes on wages. For example, supervisory and administrative are usually considered to be indirect costs of employment. Also 'hidden' costs.
in-house call centres	A call centre that is part of a larger enterprise and provides services for that enterprise.
intervention	A proposed change in the workplace or work systems to ensure, for example, better and safer working conditions and/or increased productivity. Also known as a 'proposal'.

key performance indicators	Indicators by which performance is judged. Can be applied to individuals or groups and may be quantitative or qualitative.
knowledge management	A system to record and make accessible the collective knowledge of an organisation.
KPI	*See* **key performance indicators.**
labour	In the economic sense it is used to encompass all forms of human endeavour for which there is payment including full- and part-time employees, contractors, casual employees, etc.
labour turnover	The rate at which employees leave or join a company. For example, in a company that employs ten people if two leave in any one year the turnover rate is 20% per year.
leading hand	A worker with supervisory duties but who also works, or may work, in the same tasks as those supervised.
long-term models	Due to the length of time that an intervention or proposal will take, and the consequent costs and/or benefits involved, it is assumed that the value of the money will change significantly during that period. *See also* **discount rate.**
Luddite	One who opposes the introduction of new ideas, machinery, systems. Named after English workmen (early 19th century) who organised to destroy manufacturing machinery which, they believed, was destroying their employment.
macroeconomic	The economic system beyond the enterprise level; at the national or international level.
manual lifting tasks	The application of human effort, to lift or move an object.
market	Interaction between buyers and sellers of products and services to determine the quantity of product that will be sold and the price.
market failure	Where the general assumptions of economics do not hold true, the market is considered to have 'failed'.
market price	The price of goods or services when sold in the 'market'. *See also* **market.**
material safety data sheets	Information, usually accompanying a chemical substance, that describes its contents, properties and hazards. Includes safety precautions, medical and first-aid advice. Also MSDS.
mechanical lifting equipment	Any one of a large range of devices including cranes, fork-lift trucks, patient lifters, etc. used to reduce or replace manual lifting tasks.
microeconomic	Economics at a local level, especially at the enterprise or work station level. The Productivity Assessment Tool is an example of a microeconomic model.
MSDS	*See* **material safety data sheets.**

multi-skilling   Workers trained in several workplace-specific tasks or skills, frequently to enable rotation of tasks or jobs.

musculoskeletal injury Injury to the musculoskeletal system which includes bones, tendons, joints, ligaments, muscles and nerves. Injuries are most often to the back, arms and shoulders and, if work-related, may be due to force, repetition, posture, frequency and/or duration. *See also* **cumulative trauma disorder, repetition strain injury, occupational overuse syndrome, cervicobrachial disorder.**

negative feedback   Returning a signal or an output of a system in order to restore that system to its normal milieu or status.

neoclassical economics School of economic thought in which general economic assumptions are spelled out.

net benefit   Any situation where the costs and benefits of a proposal or intervention are compared and where the benefits outweigh the costs. *See also* **cost-benefit, cost-effectiveness.**

net present value (NPV) The present value of a project's future benefits minus the initial costs. The values for benefits can be discounted and/or values for costs can be increased to account for the change in the value of money over the life of the project.

noise-induced hearing loss Deafness induced by working in a noisy environment, usually over a relatively long period.

occupational overuse syndrome A work-related musculoskeletal injury. *See* **musculoskeletal injury.**

OOS   *See* **occupational overuse syndrome.**

operator   In a call centre, those employees who make or receive telephone calls; *see* **call centre operator.**

outsourced call centres A specialist enterprise that acts on behalf of clients who are not part of that enterprise. *See also* **in-house call centres.**

overhead costs   *See* **indirect costs.**

pay-back period   The time taken to repay the expenditure for an intervention.

perfect competition   Theoretical economic concept. It assumes that there is a market in which no participant (seller or buyer) can influence prices. It is characterised by a free flow of information and a large number of buyers and sellers, with no barriers to more buyers or sellers entering this market.

precarious work/workers Applies to casual, part-time, hourly paid and/or short-term contract workers. Usually they do not have the benefits of full-time employees such as paid sick and vacation leave. Many self-employed people (including authors and ergonomics consultants) come into this category. *See* **casual employee.**

price distortions   Situations where the price of products or services differs from a perfectly competitive market, e.g. where tariffs or

	other taxes make imported goods more expensive relative to local products or where subsidies support the production costs of local products so they can be sold more cheaply.
productive employment cost	The cost of the time worked and paid for. It is the direct and indirect cost of employment divided by the paid hours actually worked (productive time).
productive time	The total paid hours actually worked. It is the paid time less hours not worked but paid for, e.g. sickness absence, vacation, training.
Productivity Assessment Tool	A short-term, cost-benefit analysis model used to assess the financial returns for workplace interventions.
products	Anything sold by an enterprise, whether physical items or services.
proposal	*See* intervention.
public goods	Factors that provide benefits to all persons and enterprises and which no one enterprise pays for, e.g. provision and maintenance of roads.
quick keys	On a keyboard, two or more keys used in combination to perform a specific function.
reduced productivity	The reduction from the optimum output at a work station or workplace. This may be due to one or more factors including lack of operator skills, unsuitable hand tools, poorly maintained capital equipment.
repetition strain injury	Work-related musculoskeletal injury. *See* **musculoskeletal injury**.
resources	Components used for producing products and services. For ease of analysis these are grouped into broad categories, commonly: human resources, natural resources, money resources, products and services.
return on investment	A method for assessing a proposal in financial terms. It considers the cost of a proposal, often termed an outlay or investment, and then calculates the benefits from the proposal taking into account the time frame which it will take to recover the initial outlay or investment. Also termed **financial appraisal**.
Roben's legislation	The UK legislation that established broad standards, decision-making structures and procedures rather than an alternative system of directly specifying detailed standards that might apply to a particular workplace, machine or work process.
ROI	*See* **return on investment**.
RSI	*See* **repetition strain injury**.
service work	Formerly white-collar work but now more generally applied to industries such as hospitality, health care, office and professional work.
shift work	All work schedules outside daylight hours.
shortcut keys	*See* **quick keys**.

short-term model	Assumes that the value of the money used to fund an intervention or proposal will remain approximately the same over the course of the project. The **Productivity Assessment Tool** is an example of a short-term model.
sitting with Nelly	An on-the-job learning process whereby a new employee works alongside an experienced employee.
social taxes	In some countries a tax placed on wages, paid for by the employer, which is designed to encompass the cost for employee injuries, ill health and retirement.
speed keys	*See* **quick keys.**
stakeholders	Those who have a financial or other interest in an enterprise. These include employees and shareholders.
static posture	A position of the limbs or body such that those muscles which hold the posture are required to work. *See* **musculoskeletal injury.**
supply	The willingness and ability of potential sellers to offer products or services at a variety of alternative prices during a particular time period.
Taylorism	The reduction of a complex process or operation to its most simple components or tasks so that each task may be undertaken by a person with little or no skill or special training. Salient features include short cycle times/high repetition of the same tasks, monotony and lack of intellectual stimulation, lack of control/responsibility over one's work. Also known as scientific management and job simplification. Named after F.W. Taylor, who developed the management theory in the early part of the 20th century.
unplanned absence	Encompasses sickness and other absences that cannot be forecast.
upper limb injury	Usually a musculoskeletal injury to the hands, wrists, arms and/or shoulders.
wrap-up time	For a call centre operator, a short period between calls.

Index

Note: page numbers in *italics* refer to figures, tables and boxes.